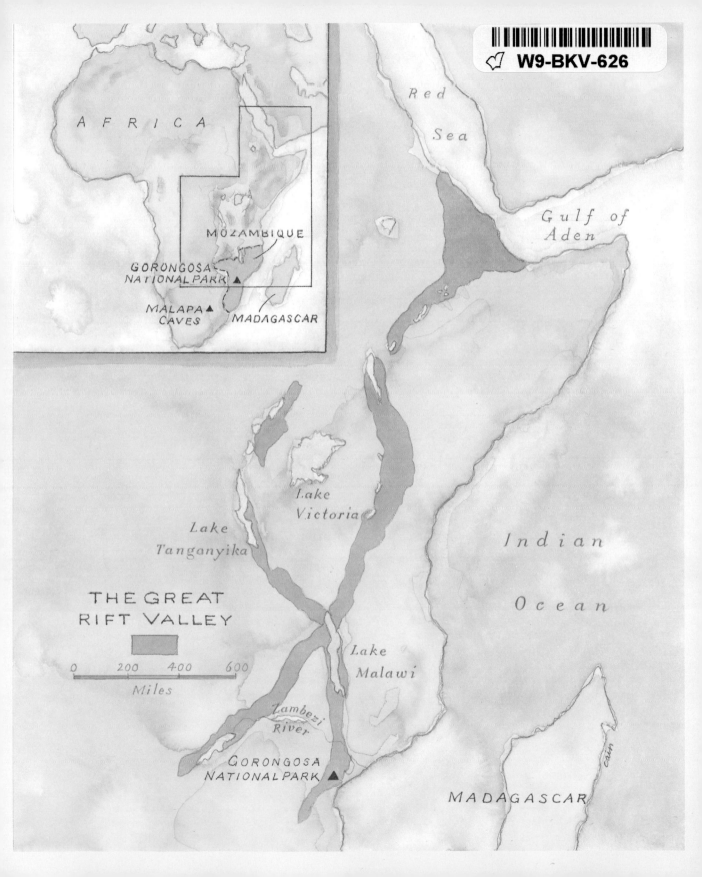

AFRICA

Red Sea

Gulf of Aden

MOZAMBIQUE

GORONGOSA NATIONAL PARK ▲

MALAPA CAVES ▲

MADAGASCAR

Lake Victoria

Lake Tanganyika

Indian Ocean

THE GREAT RIFT VALLEY

0   200   400   600
Miles

Lake Malawi

Zambezi River

GORONGOSA NATIONAL PARK ▲

MADAGASCAR

Also by Edward O. Wilson

Letters to a Young Scientist

Why We Are Here: Mobile and the Spirit of a Southern City
(with Alex Harris)

The Social Conquest of Earth

Kingdom of Ants: José Celestino Mutis and the Dawn of Natural History in the New World
(with José M. Gómez Durán)

The Leafcutter Ants: Civilization by Instinct
(with Bert Hölldobler)

Anthill: A Novel

The Superorganism: The Beauty, Elegance and Strangeness of Insect Societies
(with Bert Hölldobler)

The Creation: An Appeal to Save Life on Earth

Nature Revealed: Selected Writings, 1949–2006

From So Simple a Beginning: The Four Great Books of Darwin,
edited with introductions

Pheidole *in the New World: A Dominant, Hyperdiverse Ant Genus*

The Future of Life

Biological Diversity: The Oldest Human Heritage

Consilience: The Unity of Knowledge

In Search of Nature

A WINDOW ON
ETERNITY

A BIOLOGIST'S WALK THROUGH
GORONGOSA NATIONAL PARK

Edward O. Wilson

Photographs by Piotr Naskrecki

Simon & Schuster
New York   London   Toronto   Sydney   New Delhi

Simon & Schuster
1230 Avenue of the Americas
New York, NY 10020

First Simon & Schuster hardcover edition April 2014

SIMON & SCHUSTER and colophon are registered trademarks of Simon & Schuster, Inc.

For information about special discounts for bulk purchases, please contact Simon & Schuster Special Sales at 1-866-506-1949 or business@simonandschuster.com.

The Simon & Schuster Speakers Bureau can bring authors to your live event. For more information or to book an event contact the Simon & Schuster Speakers Bureau at 1-866-248-3049 or visit our website at www.simonspeakers.com.

Interior design by Nancy Singer
Jacket design by Marlyn Dantes
Jacket photographs by Piotr Naskrecki

Manufactured in the United States of America

1 3 5 7 9 10 8 6 4 2

Library of Congress Cataloging-in-Publication Data
Wilson, Edward O.
A window on eternity : Gorongosa National Park, Mozambique / Edward O. Wilson; photographs by Piotr Naskrecki.
Includes index.
1. Biodiversity—Mozambique—Parque Nacional da Gorongosa. 2. Natural history—Mozambique—Parque Nacional da Gorongosa. 3. Restoration ecology—Mozambique—Parque Nacional da Gorongosa. 4. Nature conservation—Mozambique—Parque Nacional da Gorongosa. 5. Parque Nacional da Gorongosa (Mozambique)—Environmental conditions. 6. Parque Nacional da Gorongosa (Mozambique)—Description and travel. 7. Parque Nacional da Gorongosa (Mozambique)—Pictorial works.
I. Naskrecki, Piotr. II. Title.
QH195.M6W55 2014
333.95'1609679—dc23
2013032607

ISBN 978-1-4767-4741-5
ISBN 978-1-4767-4743-9 (ebook)

For Gregory C. Carr,

world citizen, who conceived and brought to reality

the rebirth of Gorongosa National Park

I have set before you life and death,

blessing and cursing: therefore choose life,

that both thou and thy seed may live.

—Moses at Mount Nebo, Deuteronomy 30:19

CONTENTS

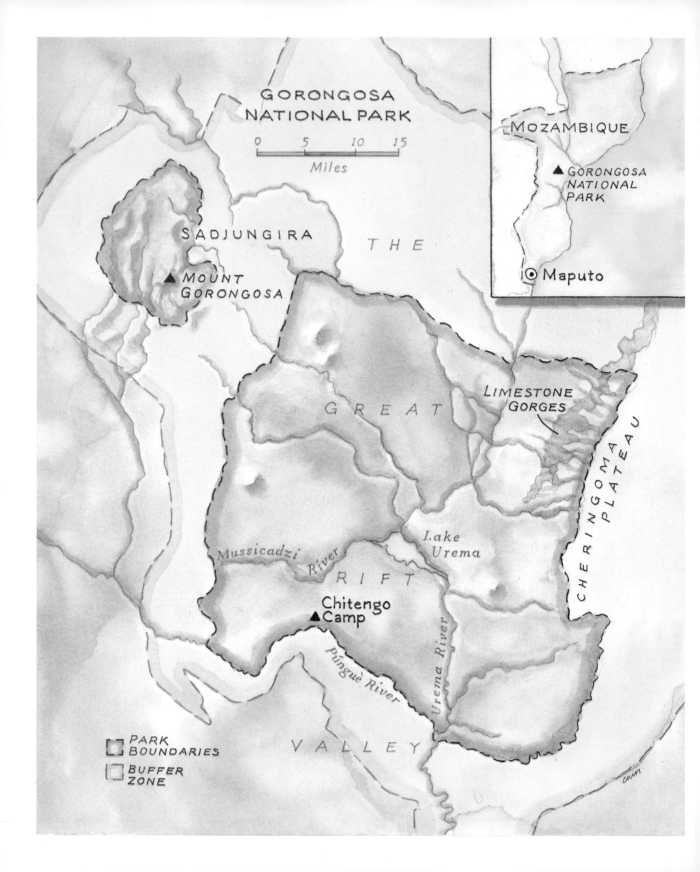

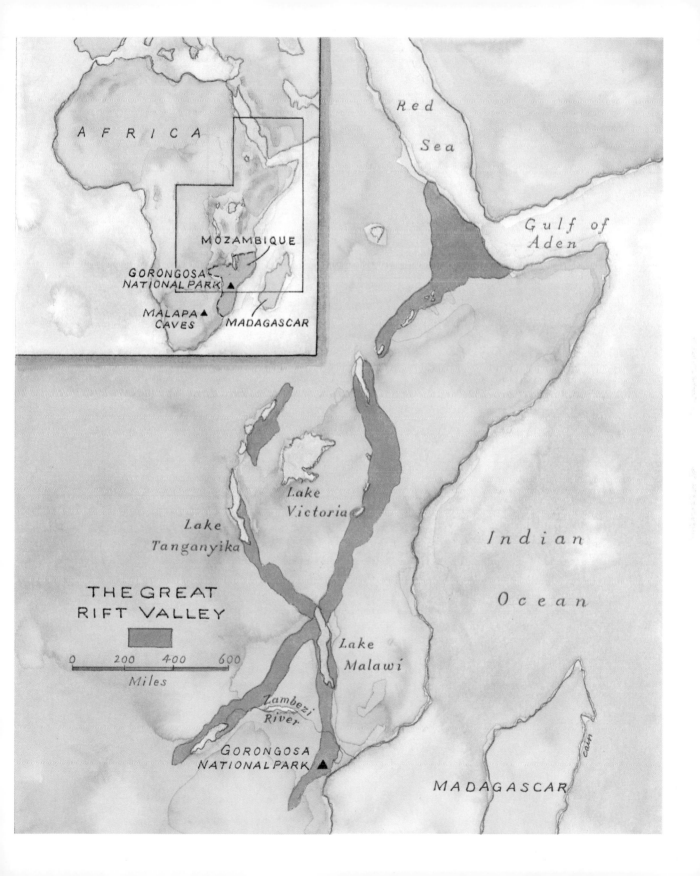

AFRICA

MOZAMBIQUE

GORONGOSA
NATIONAL PARK ▲

MALAPA ▲
CAVES

MADAGASCAR

Red
Sea

Gulf of
Aden

Lake
Victoria

Lake
Tanganyika

THE GREAT
RIFT VALLEY

0   200   400   600
Miles

Indian

Ocean

Lake
Malawi

Zambezi
River

GORONGOSA
NATIONAL PARK ▲

MADAGASCAR

cain

A WINDOW ON
ETERNITY

THE SEARCH FOR ETERNITY

Oblivion, absolute oblivion, is the one image the human mind cannot accept or even fully conceive. Deeper than despair, more terrifying than death, is the thought that everything in time will disappear, that all we have been and will become will leave no trace whatsoever. We accept that the body must die and disintegrate. Physical death is not, however, the same as true oblivion. There remains the hope of eternal life in some other, astral dimension. And if that possibility is abandoned, then left behind are memories in the minds of a still-living family and their descendants to come. When these too dissipate, as surely they will, there remains the tribe of which the self was part, and beyond, there is the nation. Still, these too must come to an end. Across millennia and tens of millennia, the mill of time grinds all remembrance into nothing.

Is there any shield against the Dismal Thought except just to shut it from the mind or not bring it up at all? Well, there is the whole human

● ● ●

During the dry season there is little water in the atmosphere, which makes for star-studded skies and very cold nights in Gorongosa National Park.

species, and the role we play in particulate human heredity. The body of each person contains the DNA bequeathed by two parents, four grandparents, eight great-grandparents, and so on back in time, the number doubling every generation. Each person has as many as five thousand ancestors who lived only three centuries ago. Looking to the future, at least a large percentage of your genes will be spread—if not by direct descendants then through collateral relatives, cousins to one another in each slice of time.

In addition to this enduring community of DNA, there is culture. The grand ensemble of our personal genes will dissolve into endless new combinations, but culture will evolve and live on in more or less coherent form. Because of the common blood and remembrance of all humanity, oblivion can be put off forever—might it not?

Perhaps. Yet even the reprieve from oblivion by membership in the human race is not guaranteed. All species in the history of life have been vulnerable to extinction. Since the origin of life on Earth, about 3.5 billion years ago, 99 percent of those species that ever existed have come to an end, to be replaced by other species. Species multiply as they evolve and they eventually decline, to be replaced by other arrays of multiplying species that also come to an end. Worse, *Homo sapiens* is a kind of mammal, the species of which are statistically more vulnerable to extinction than those of other kinds of animals. From the fossil record, scientists have found that mammalian species enjoy an average longevity of only about half a million years, a relatively short period in the living world as a whole. Evolving populations reached the *Homo sapiens* level of physical traits and probably also full mental capacity about 100,000 years ago. Humankind was then the survivor of prehuman species that had lived and died during the hundreds of millennia of our ancestry. Granted that our unprecedented culture and rational power may vouchsafe humanity a much longer life span. Yet, maybe not. Our unique powers could well shorten humanity's time on Earth.

Even so, there is another way, logical and moral in substance, to reject the inevitability of oblivion. We should not think that all our ancestors and all that we are today must someday come to nothing. It is neither too early, nor the topic too remote from day-to-day reality, to envision a different kind of immortality, and how it might be achieved. It resides in those remnants of the natural world we have not yet destroyed. The rest of life is a parallel world. It could exist and continue evolving for what to the human mind is an eternity.

Let me explain.

•••

A bird's-eye view of the vast expanse
of Gorongosa as seen by the author
from the cockpit of a helicopter.

one

THE SACRED MOUNTAIN
OF MOZAMBIQUE

In the summer monsoon season of late November to mid-March the rain clouds ride on the trade winds of the Indian Ocean into the center of Mozambique. Crossing the coast, they refresh the miombo dry forest of the Cheringoma Plateau, then the savanna and floodplain grasslands of the Great Rift Valley. They run aground on the windward slopes of Mount Gorongosa, where, like a benediction to the fertility of Africa, they release great torrents of rain.

It was in the dry season of 2011 that I arrived at the small coastal city of Beira and first met Greg Carr. We immediately boarded a helicopter and flew inland across the Cheringoma Plateau to Gorongosa National Park. It was a formative experience for me. Although I had traveled around the world on field expeditions, this was my first visit to Africa south of the Sahara—the *real* Africa for many. I was awed as any first-time tourist by the spectacle of the big animals—elephants, hippopotamuses, crocodiles, and herds of diverse antelopes—as we flew as unobtrusively as possible above them. Here, I reflected, was the last megafauna on Earth left more or less intact to roam open savanna and grassland.

Gregory C. Carr is an American businessman and philanthropist who has adopted the park and its fauna and flora as his lifetime mission. By my second visit, a year later, I had begun to conduct and promote research on the previously little-known insect fauna. Greg and I, bound by a common philosophy and broad overlap in the science of wildlife conservation, were by then close collaborators and fast friends.

Greg knew from the start that Mozambique's sacred mountain, better called a massif because of its several peaks, reaching a height of 6,112 feet (1,863 meters), is the heart of the park. It is high enough to capture two meters (79 inches) of annual rainfall and support a lush rain forest on the summit. The latter habitat, about 18,500 acres (75 square kilometers) in extent, is an ecological island in a sea of savanna and grassland. Because it is both remote and easily accessible only by helicopter, the forest has remained largely unexplored by biologists. Ants, my own specialty, were entirely a blank on the map when I first came to Gorongosa in 2011. For a naturalist there can be no more powerful magnet than an unexplored island, or the equivalent of an island like this island mountain peak, with a little-known fauna and flora awaiting study. When I visited the summit by helicopter, I was as excited as I have ever been in my long career by the prospect of surprise and discovery.

Gorongosa National Park, with the mountain forming its northwestern border, is a global treasure. Its forested plateau is the southernmost extension of Africa's Great Rift Valley, a domain of savanna, dry forest, and grassland centered around Lake Urema and fed by a lacework of rivers and tributary streams. The eastern slope of Mount Gorongosa is covered by a series of vegetation zones weakly divided by elevation and capped by the summit rain forest. Close by rise three seldom visited inselbergs, miniature peaks of lower elevation. To the east of the massif is a region of caves and exposed limestone ridges and intervening deep ravines, their sides and bot-

●●●

Looking west from the plains surrounding Lake Urema is the silhouctte of Mount Gorongosa in the distance.

The slopes of Mount Gorongosa are covered with lush, verdant rain forest, a remnant of the immense forest that once stretched across most of the African continent.

toms heavily grown in rain forest, also awaiting exploration. This mix provides the greatest variety of habitats of any park in Mozambique, and one of the most diverse in the entire world. In the whole of the park have been found so far 398 bird species (about 250 are residents), compared with 914 in all of North America; 123 mammals, of which one indigenous species is *Homo sapiens,* and which is well above the 67 species living, for example, in Yellowstone National Park; 34 reptiles; and 43 amphibians. It is just an educated guess, but I am reasonably certain that the resident species of insects, arachnids, and other invertebrates number in the tens of thousands.

These raw figures are impressive, but they cannot capture the aesthetic impact of the fauna and flora. I prefer to leave that task to Kenneth

• • •

Mount Gorongosa is the source of four major rivers, which provide water to Gorongosa and all communities surrounding the mountain. Some form spectacular waterfalls on its slopes, the largest being the Murombodzi Falls, more than one hundred meters (328 feet) high.

Tinley, who pioneered the natural history of the park in 1968–73. "The valley mists," he said of an early winter morning in the Rift, "envelop everything in an ethereal world made mystic by the yellow trunks of fever trees and parasol shapes of palms. The sweet scent of acacia flowers or the raw-green smell of the potato-bush infuses the air, the impala look like they are walking on air, and baboons wait in the canopy of large winter thorn trees to catch the warmth of the rising sun."

Gorongosa in the Mwani tongue means "place of danger." And it

● ● ●

The Mount Gorongosa pygmy chameleon (*Rhampholeon gorongosa*) was discovered only in 1971. This and many other animals and plants are species endemic to the mountain. Here a very young individual has climbed the head of an adult female. While this behavior was likely accidental, males of this species often ride on females' backs in an effort to discourage rivals.

has been just that. Gorongosa is today still recovering from a tragic recent history. After Mozambique won its independence from Portugal, in 1975, a civil war broke out and raged for sixteen years. Because the park was close to the headquarters of Renamo (Resistência Nacional Moçambicana), one of the opposing armies, it became a battleground. Its tourist facilities were destroyed. Roving soldiers, hungry for any food they could forage, killed many of the large animals that once teemed in the park. They focused on elephants, for their meat and ivory could be traded for weapons in South Africa. After national peace accords were signed, but before order could be restored at Gorongosa, commercial poachers killed an even larger number of animals and peddled their meat at nearby markets. At the end, gone or pushed close to extinction were the elephants, Cape buffaloes, hippopotamuses, lions, warthogs, and more than a dozen species of antelopes. Only the crocodiles, quick to slide down muddy banks into the waters of the Púnguè, Mussicadzi, and Urema rivers, and thereby difficult to kill and retrieve, escaped with little harm.

The temporary clearance of big game had important environmental consequences. Where zebra herds no longer grazed, grass and woody shrubs thickened into tinder, and lightning-struck wildfires became more threatening. Where no more elephants knocked over trees in order to feed off their branches, some forests increased in density. With the scat and carcasses of big game severely reduced, the population of scavenging insects and other animals must have fallen sharply. Yet the platform of vegetation and small animals, including the myriad species of insects and other invertebrates, remained largely intact. I don't have enough data to support my first impression, but I believe the lower levels of the food chains held fast, awaiting the return of the megafauna.

For a decade following the end of the civil war, while a new, democratic Mozambique established itself, Gorongosa National Park remained

in ruins. In 2004, Greg Carr was asked by the government of Mozambique to evaluate the condition of the park and help plan its restoration. Greg, a dynamo of ideas and action, did just that and much more: he made Gorongosa's restoration his full-time occupation. With the approval of the government and enthusiastic support of the local people, he undertook, largely at his own expense, the restoration of the park's megafauna and tourist facilities. Within less than a decade, Gorongosa was well on its way to recovery. Large animals, including Cape buffaloes and five elephants, had been imported from nearby South Africa and were multiplying rapidly. Cheetahs and African wild dogs were on the list for future restoration. While still well below their prewar maximum, herds of grazers and browsers once again swarmed across the savanna and grasslands. They were growing fast. The return to the ecological balance struck by the old megafauna was on course. Excellent facilities were in place at the central Chitengo Camp and outlier "explorer" camps in the interior. By the time of this writing (2013), the number of visitors from Europe and North America had risen to a corresponding degree.

Anyone witnessing the rebirth of Gorongosa cannot escape appreciating the magnitude of the accomplishment by Greg Carr's team and the people of Mozambique. It is one thing to draw a line around a beautiful natural area, declare it to be a national park, then add the amenities necessary to serve the public. It is entirely another thing, at a higher order of magnitude, to restore a damaged park to its original health and vibrancy.

During my first stay at the park my assistant was Tonga Torcida, a young man born on Mount Gorongosa. He was one of the first of his

• • •

The rain forest of Mount Gorongosa is home to many unusual animals and plants, such as this sylvan katydid (*Acauloplax exigua*).

village to graduate from high school—no minor feat since schooling past the seventh grade requires tuition and a uniform, affordable by only a few local families. While we worked together, Tonga learned that he would receive a scholarship to attend a Tanzanian wildlife college. In January 2013 he entered his first class. His effort is a spearhead for his people and Mozambique's natural environment. Speaking four languages and working off his intimate knowledge of the local environment, he plans to be a wildlife biologist.

Greg Carr and I decided to hold a bioblitz on Mount Gorongosa and use it to engage the community living on its lower slopes. Up to seven thousand people live on the mountain, spread out in single dwellings or tiny villages of up to several families. All (except those with jobs in the park) live on subsistence agriculture. We asked Tonga to help organize the event, and in particular to recruit local children as our helpers. Bioblitzes are counts of numbers of local species found and identified in a restricted area and during a fixed period of time, usually twenty-four hours. They follow a simple set of rules: local naturalists familiar with one or more groups of organisms are invited to assist the search and identify the species discovered by themselves and other participants. The search for species is conducted within a selected radius around the focal point. One of the earliest bioblitzes ever held was at Concord, Massachusetts, on July 4, 1998, with Walden Pond, made famous by Henry David Thoreau, as the focal location. The idea had been conceived by Peter Alden, a master naturalist and writer from Concord. On that day I was proud to work by his side. Naturalists, many of them amateur and professional specialists on different kinds of plants and animals, came from all over New England. The weather was ideal, and the excitement intense. The effort was so successful and well publicized that similar events have in time been held all over the

United States, including Central Park (twice), in addition to at least eighteen other countries.

On July 27, 2011, we held the Gorongosa event on the eastern slope of the mountain at 3,600 feet elevation, just beneath the lower fringe of rain forest. There we met children and their parents from a village located nearby on the lower slope of the mountain.

I began with a speech about the whole thing, with Tonga translating English into the local dialect. I spoke slowly, because I knew it was no easy job for Tonga.

We're all going to hold the very first bioblitz seen on this mountain, and possibly the first in all the country of Mozambique. Everyone can join. We will find all the kinds of animals we can, living in this particular place, and here and now on the eastern side of Mount Gorongosa. The game is this: how many kinds of insects and other kinds of animals can we find in two hours in this spot on the mountain? We are going to have a combination of science, social gathering, and treasure hunt. I helped out on one of the first ones, held in the United States over ten years ago. Since then, bioblitzes have been held all over the world. Usually, there are many experts to identify different kinds of animals, such as birds, fishes, insects, and spiders, but today I am going to be the only one for this job, and I will do my best. You find the creatures, and I'll sit here and identify them for you.

Why is a bioblitz important? Because what we are doing today is real science! Scientists and everybody need to know what the different kinds are in order to study and understand life here, and take care of the environment. Okay, let's begin the Mount Gorongosa bioblitz.

In this case, bending to logistic necessity, we limited the time to two hours, and I served as the sole expert. I was able to identify most of the insects and spiders to their taxonomic families (such as millipedes of the family Julidae, rove beetles of the family Staphylinidae, and of course, unerringly, ants, all belonging to the family Formicidae). For some specimens I had to make a rough guess.

The Mount Gorongosa bioblitz was almost certainly the first ever held in Mozambique. In a melee of scurrying and shouting, the children, about four to twelve years old, proved remarkably gifted hunters. They were eager to hear what I had to say about their discoveries. It became quickly obvious to me that these kids were a great deal better prepared to find wild biodiversity than their age equivalents in America and Europe, most of whom live in urban and suburban environments. Considering the limited time, their young age, the relatively poor biodiversity in which they hunted (grassland next to a mountain stream, chosen for a safe helicopter landing), and the season (the height of the winter dry season), their harvest was impressive.

Here is the final list of the insects and other animals they brought to me, identified to taxonomic order and family, with the number of species found in each family:

ODONATA (dragonflies and damselflies): Libellulidae, damselflies, 1 species.

NEUROPTERA (lacewings and relatives): Chrysopidae, lacewings, 1 species.

ORTHOPTERA (grasshoppers and relatives): Acrididae, grasshoppers, 6 species; Gryllidae, crickets, 1 species.

DICTYOPTERA (praying mantises, cockroaches, termites): Mantidae, praying mantises, 4 species.

••• Bioblitz.

PSOCOPTERA (bark lice). Psocidae, barklice, 1 species.

HEMIPTERA (true bugs): Reduviidae, assassin bugs, 5 species; Belostomatidae, giant water scorpions, 1 species; Coreidae, squash bugs, 2 species; Pentatomidae, stink bugs, 1 species; Cicadellidae, leafhoppers, 1 species.

COLEOPTERA (beetles): Tenebrionidae, darkling beetles, 1 species; Silvanidae, flat grain beetles, 1 species; Coccinellidae, ladybird beetles, 1 species; Carabidae, ground beetles, 1 species; Cerambycidae, longhorn beetles, 1 species; Staphylinidae, rove beetles, 2 species; Curculionidae, weevils, 1 species; Gyrinidae, whirligig beetles, 1 species; unidentified, 1 species.

DIPTERA (flies): Tipulidae, crane flies, 1 species; Drosophilidae, vinegar flies, 1 species.

HYMENOPTERA (sawflies, bees, wasps, ants): Formicidae, ants, 3 species; Ichneumonidae, ichneumon wasps, 1 species; Vespidae, paper wasps, 1 species.

LEPIDOPTERA (butterflies, moths): Lycaenidae, blues, 1 species; Pieridae, whites, 3 species.

MYRIAPODA (millipedes, centipedes): Julidae, julid millipedes, 1 species.

CRUSTACEA (crabs, shrimp, pill bugs): Potamonautidae, freshwater crabs, 1 species.

ARANEA (spiders): Thomisidae, crab spiders, 1 species; Heteropodidae, huntsmen spiders, 1 species; Salticidae, jumping spiders, 1 species; Tetragnathidae, long-jawed spiders, 1 species; Araneidae, orb weaving spiders, 2 species; Linyphiidae, sheetweb weavers, 1 species; Lycosidae, wolf spiders, 1 species.

AMPHIBIANS (frogs, salamanders): Ranidae, true frogs, 1 species.

REPTILES (snakes, lizards, turtles, crocodilians): Agamidae, agamid lizards, 1 species.

AVES (birds): flying in immediate area, not identified. 3 species.

MAMMALIA (mammals): Muridae, mice, 1 species.

The total for the two-hour search by thirty children was 61 species belonging to 38 taxonomic families of animals, almost entirely insects. This compares quite favorably with the 777 species of animals reported in the 1998 Concord, Massachusetts, bioblitz, when you consider that that one continued over a period of twelve hours by a large crowd of adults and children, including several dozen amateur and professional specialists, over a much larger area and in July, the height of the summer season.

During the civil war of 1976–92, as marauding soldiers invaded the mountain, subsistence farmers began to open little plots higher and higher up the slope. The inviolability of the sacred mountain, a place still regarded as the birthplace of humanity by local people, was largely either forgotten or ignored. Eventually the incursion reached the summit rain forest; tall trees were felled and the moist, fertile soil converted into corn and potato fields. From 2005 to 2011, the area of original rain forest was reduced by almost half.

It was clear to ecologists and wildlife managers at the time I saw the damage, in 2011, that the clearing of the summit forest would have severe consequences if not quickly halted. The reduction of its area already meant that fewer plants and animals, some found only in this one place, can continue to live sustainably. The complete removal of the forest, which if unabated might easily have occurred within ten years, would have been catastrophic for the entire park. The mountain's ability to capture, hold, and gradually release monsoon rainwater would be gone. The water would then run off the mountain quickly, and the moisture supplied to the rest of the park rendered seasonal instead of continuous year-round. In the face of the new aridity, life in and around the park would be less sustainable for both wildlife and people.

In 2010, after several years of lobbying, Greg Carr and his Mozambique allies persuaded the government to add Mount Gorongosa to the park boundaries, an omission that had dated to the first setting of the boundary in 1960. The decree became effective in 2011, giving park officials authority to establish security around the forest perimeter. Carr then hired teams to create farms of nurseries to grow seedlings of the rain forest trees, and to begin the decades-long, perhaps centuries-long, process of returning the forest to its original area. Gorongosa National Park is creating schools and health clinics for local people at the base of the mountain below the rain forest, integrating human development with conservation. Finally, a center for scientific research and education has been built at Chitengo Camp, located at the southern boundary of the park close to the Púnguè River. The center will open in the spring of 2014. The emphasis of the research is on the natural environment in the park and the preservation of its biodiversity. The local people understand. They and their children are the beneficiaries. Here, then, in one of the remotest parts of Africa, a great environmental tragedy has been averted just in time.

two

ONCE THERE WERE GIANTS

Beyond reasonable doubt, the human race was born in Africa. However, in and around Gorongosa two very different stories are told of how this momentous event occurred. The more venerable of the two, passed from generation to generation among the indigenous people, identifies the sacred mountain as the birthplace of our species in its present form. As told by Jordani Dique at Sadjungira, a village at the mountain base, in 2009, humans were once giants who dwelled in close communion with God on and around the sacred massif of Gorongosa.[1] The strongest among them was Kangamyi, who lived with his wife on the highest point of the mountains. He could travel thirty miles in one step—to Bunga, for example, a mountain village on the Zimbabwe border. Kangamyi left a huge footprint on a rock at the summit, where it can be seen today. The giants were well disposed, but presumptuous to a fault. They asked a great deal of God, and eventually went too far, provoking his annoyance. What happened then was told to Jordani Dique by his father.

●●●

Gorongosa's caves were once home to our ancestors.

God lived on top of the water of the Sea. He used to live with us on Earth. But we bothered him too much. Imagine God is here in Sadjungira and there is a drought. Soon everyone is coming to him, asking for water. So when this happened, God thought, "Everyone is bothering me too much," and he fled to Heaven. But people thought, "Let us follow him and ask him for what we need." They began to climb one on top of the other, standing on each other's shoulders, trying to reach Heaven. When God saw this he thought, "Oh no. These men are trying to reach me even here. They can do this because they are giants. I will make them short." And that is what he did.

The second creation story is the one being created by science, century by century in the retracing, archaeological site by site in the location, and gene by gene along the twenty-three pairs of human chromosomes. In one respect the scientists agree with the poets of Gorongosa: the right place is indeed Africa. Looked at from a purely zoological viewpoint, the people of the mountain and all humans in the world are as much part of the indigenous fauna of this continent as its lions, elephants, and antelopes.

If *Homo sapiens* was to be born anywhere, the mammal fauna and environment predisposed it to be here. For tens of millions of years, Africa has always been the homeland of large primates that spend a large part of their life in the group away from trees. Five species of baboons, which are big, dog-headed monkeys with short, opposable thumbs, hunt for food on the ground during the day and roost in trees and on cliff faces at night. Chimpanzees, bonobos, and gorillas, humanity's closest genetic relatives, live by basically the same routine. Vervets are the most widespread and locally abundant of all of the continent's monkeys and apes. In Mozambique, they are distinguished by reddish green body fur,

●●●

The dominant primate
species of Gorongosa
is the highly intelligent
and always curious
yellow baboon.

●●●

Vervet monkeys are
mostly vegetarian and
are frequently seen
feeding on trees in
bloom.

black faces, white foreheads, and pink eyelids. An adult male, the larger sex, weighing as much as twenty pounds, is instantly recognizable by its pale blue scrotum and bright red penis, used in the "red, white, and blue" display during dominance conflicts with other males. Vervets move easily through both trees and over the ground, in troops of 8 to 140.

Three species of larger mammals regularly venture onto the grounds of Chitengo Camp of Gorongosa: warthogs, olive baboons, and vervets. The warthogs, often kneeling to get closer to the ground, clip grass and other low vegetation. They also dig through the lawn soil for roots, making serious work for the staff gardeners. The two primates are omnivores that feed opportunistically. At Chitengo Camp, they use their

• • •

Kneeling to get closer to the ground, warthogs are a common sight in the grasslands of Gorongosa.

fingers and hands to pick insects, seeds, berries, grass, and other bits of edible vegetation as food.

The unusual willingness to mingle with humans at Chitengo Camp almost certainly arises from greater intelligence. Apes and monkeys score highest in intelligence tests among terrestrial animals. Not far below are the swine, composing the taxonomic family Suidae, whose African species include the bushpigs and warthogs. Rivaling the swine are the three species of elephants (yes, there are three, not two as you might suppose: the savanna and forest elephants of Africa and the Indian elephant of Asia). A bit down in rank are dogs, then much farther down are domestic cats.

Might it be that the smartest animals learn more quickly how to deal with humans, in clever and opportunistic ways? At Gorongosa this appears to be the case. The baboons and vervets work their way cautiously out of the forest and into Chitengo Camp like soldiers on patrol in enemy territory, with a few venturing cautiously here and there, and others following in sight of them. These smart primates are not easily repulsed. Where impala and other antelopes freeze and stand quietly at the sight of a human, ready to bolt suddenly and run away, baboons and vervets are more likely to retreat only a short distance and then to one side, halt, turn, and watch for a while. If the human moves on past, the alert animal soon resumes its search for food. If approached directly, it angles away again, maintaining a safe distance, watching, looking about, and—it is hard to avoid the conclusion—*thinking* about the situation. In any case, it does not flee from the area unless directly chased. It wastes little of its feeding time managing humans. There are also bush babies (also primates), and I have been told that once in a great while a lion strolls by, but otherwise I have never seen another kind of large mammal on the grounds of Chitengo Camp.

As I walked around and close to these smart primates, which were minding their own business and wishing, I suppose, I would do the same, I could not help but think about humanity's distant ancestors in the same habitat, perhaps in the very same spot, then forest-covered, searching for food while watching for lions, leopards, and other predators. As big mammals foraging over the ground with pulpy fingers and flat nails, and neither fangs nor claws with which to stab and tear their enemies, they had evolved to depend heavily upon the use of intelligence.

Six million years ago the stage had been set by big mammals in the seasonal savannas and woodlands of Africa. It was in this period that the epic of human origins began. The key step was the division of a single species of large primates into two daughter species. One was ultimately to produce the modern-day chimpanzees and their close relatives the bonobos. The other lineage led to us.

The evolutionary pathway to humanity was not a straight line. It was a crooked walk by a species through what can be usefully envisioned as an endless maze. As the environment changed, so did the ensembles of genes adapted to it. Each genetic shift took one path as opposed to others, opening the species to some future pathways while closing it to others. From time to time, an accumulation of changes caused the species to divide into two or more species.

Evolutionary biologists, by studying animals and plants of many kinds, have learned a great deal about this multiplication process, which they call speciation. The process typically begins by the physical separation of a single population into two or more populations, which then occupy different geographical ranges. The division can occur when individuals from the ancestral range occupy a new area, as when colonists reach a previously unoccupied island or valley. The division can also occur when the environment within the range changes, making parts of it uninhabitable for the species, thus breaking the species into multiple, isolated populations. As a result of their isolation, the newly created subpopulations adapt to the local conditions in which they live, causing them to diverge genetically from one another. An example of a primate in this early stage of speciation is the vervet, with three races (or subspecies, as taxonomists classify them) based on differences in color of their pelage. Experts place all three in the same species, *Cercopithecus aethiops,* but they are distinctive enough to have picked up the following common names: "typical" vervets range from southern Ethiopia through Mozambique to South Africa; grivets occupy parts of Ethiopia; and green monkeys occur from Senegal to Ghana.

With time, such populations may diverge so far in heredity that wherever and whenever they meet in the wild, they do not freely interbreed with one another. The failure may occur because their mating behaviors do not mesh, or they mate in different seasons or at a different time of day, or, if mating does happen, the offspring die early. Or finally, like mules, the offspring of horses and donkeys, they are sterile at maturity. If any of these incompatibilities occur in nature, the populations are now full species. The original, mother species is said to have multiplied into two or more daughter species. The new entities are called sister species to one another. For example, the populations that spawned the chimpanzee and human lines, respectively, six million years ago were sister species to one another.

Such newly formed species are destined to diverge further from one another. They are also likely to specialize to prefer different habitats or to use different kinds of food or forage at different times of the day or any combination of these. When they have drifted apart far enough, they can avoid competition if by chance they meet. At this point their geographic ranges may in fact spread and overlap, so that in some places two or more species occur where once in geological time there was only one. By comparing their DNA, evolutionary biologists can determine whether they are sister species. Using DNA and protein evidence, combined with knowledge of geological and climatic change, the researchers can reconstruct the history of the speciation event.

Repeated speciation events often lead to an accumulation of many species in the same area, with each specialized to a different way of life. In biodiverse parts of the world, such as the rain forests and Great Rift Valley savannas of Africa, and on large, distant islands, it is common to find large assemblages of species with recent common ancestry but displaying a range of adaptations in anatomy, physiology, and behavior. The total process by which this pattern is created is called adaptive radiation. When the number of species is large, the assembly is called a species flock.

A classic example of adaptive radiation that created an impressive species flock is provided by the primates of Madagascar, the giant island miles off the coast of Mozambique. All of the 33 known living species of primates there (at least 15 others have been driven to extinction by human activity) are lemurs, a major group distinct from the monkeys and apes of mainland Africa. All appear to have descended during millions of years from one or a very few species of primitive primates that made their way, likely by floating debris from the African coast across the several-hundred-mile widths of the Mozambique Channel, to Madagascar. No other primates were present on the island upon their arrival. Humans became the next addition, when some two thousand years ago the first voyagers arrived from Indonesia. Because of Madagascar's isolation, very few other mammals made landfall as well. Madagascar thus offered many empty niches—mostly comprising combinations of habitat and food—for the ancestral species of lemurs to squeeze in and fill. Today the surviving forms range from tiny mouse lemurs to approximately cat-sized indris, sifakas, and ring-tailed lemurs. They include the amazing aye-aye, a ghoulish nocturnal and bat-eared animal that hangs upside down on trees and hooks out beetle grubs with its extremely thin, elongated middle fingers.

We now return to the human story. Another adaptive radiation, even grander than that of Madagascar and initiated tens of millions of years ago, was achieved through Africa and Asia by cercopithecoid Old World monkeys and great apes. Today about half of the eighty species live in Africa. From their midst six thousand millennia ago arose the common ancestors of humanity and the two species of chimpanzees.

No later than 4.4 million years before the present, one of the species, *Ardipithecus ramidus*, developed an anatomical innovation that was crucial for the changes that created humanity. Although the size of its brain remained close to that of the modern chimpanzee, the hind legs of *Ardipithecus* were elongated and suitable for walking erect. Chimpanzees can stand on their hind legs while raising their arms. They can hobble along for a while, but when traveling on the ground they add balance and speed by knuckle-walking—dropping their fists to the ground and using all four limbs to travel with the same basic locomotion as a dog or other quadruped. *Ardipithecus*, in contrast, walked more or less straight up. It also retained long arms suitable for partial life in the trees.

By four million years before the present, *Ardipithecus* or a close relative had given rise to a group of species called the australopithecines (after one of the first fossils found and formally named *Australopithecus*). These prehumans took the trend to bipedal walking much further. Their body as a whole changed to accommodate efficient bipedal locomotion, as follows. The legs were lengthened and straightened, and the feet elongated to produce a rocking movement as the prehumans walked and ran. And the pelvis was refashioned into a shallow bowl to support the viscera, which now pressed downward toward the legs instead of being hung apelike beneath the horizontal body.

Bipedal, straight-up walking, the hallmark of the australopithecines, was probably responsible for the success they enjoyed. They underwent a modest adaptive radiation, expressed anatomically among their species by differences in body form, jaw musculature, and dentition. They were evidently mostly or all vegetarian, feeding on fruit, seeds, tubers, and other vegetation, with probably an occasional small animal or piece of larger animal carcass retrieved in competition with other scavengers. The details of all this we do not know, but the skull structures suggest that the species were specialists on different kinds and degrees of toughness of plant food. Those with heavy jaws presumably consumed harder materials, such as palm nuts, and those with more slender mandibles can be assumed more likely to have favored berries and other softer fruit. A few added some animal material to their diet, as mentioned.

About two million years ago, at least three australopithecine species strode across the savannas and dry tropical forests of the African continent. At a distance and a bit out of focus, they would have resembled modern humans. They almost certainly lived in small groups, in the manner of the early humans *Homo erectus* and modern hunter-gatherers. For primates whose bodies had been originally crafted for life in the trees, the bipeds could run swiftly. But they could not match the four-legged animals they hunted as prey. Antelopes, zebras, ostriches, and other animals were able to outrun them with ease over short distances. Millions of years of pursuit by lions and other carnivore sprinters had turned prey species into hundred-meter champions. Yet, if the early humans could not outsprint such animal Olympians, they could at least outlast them in a marathon. At some point humans became long-distance runners. They needed only to commence a chase and track the prey for mile after mile until it was exhausted and could be overtaken. The prehuman body, thrusting itself off the ball of the foot with each step and holding a steady pace, evolved a high aerobic capacity. In time the body also shed all of its hair, except on the head and pubis and in the pheromone-producing armpits. It added sweat glands everywhere, allowing increased rapid cooling of the naked body surface.

In the later stage of the australopithecine adaptive radiation, one species took a turn in evolution that foreshadowed the final advance to modern humanity. It shifted to a heavy reliance on meat, obtained by scavenging, hunting, or both. The change had occurred by the time of *Homo habilis,* known from fossils found in Tanzania and dated to 1.8 to 1.6 million years before the present. The innovation was accompanied, then and shortly thereafter (that is, shortly by geological standards), by the control of fire. Ground fires spreading from lightning strikes are a frequent occurrence in the savannas and grasslands of Africa today. A few animals, especially the young, sick, and old, are trapped and killed. The roaming prehumans could not have failed to discover some of the felled animals already cooked, with the flesh easily torn or cut off and ready to eat. During this period, the brain size began to grow—and in spectacular fashion. Where early australopithecine brains were between 400 and 550 cubic centimeters in volume, the *Homo habilis* brain volume was 640 cubic centimeters, half that of modern humans. That is enough, along with other changes evident in the skull, to justify classifying the species in the genus *Homo,* along with us (*Homo sapiens*), instead of the ancestral genus *Australopithecus.*

Homo habilis, or a species close to it, evolved in relatively short geological time into *Homo erectus,* with an even larger brain. Individuals of this species, the direct ancestor of *Homo sapiens,* dwelled in campsites and warmed themselves by controlled fires. Campsites, where small groups could settle and send out hunters, brought individuals intimately together for extended periods of time, as they do today. They provided a setting in which group members had to communicate almost continuously in one fashion or another, divide labor, communicate intensively, and evaluate the wishes and intentions of others. Groups able to conduct these relationships more effectively must have prevailed over those less gifted. Intelligence would have been at a premium. Brain size increased further, particularly in the cortical centers of memory, to the *Homo sapiens* level.

The rise of *Homo* out of the australopithecines, from 2.5 to 1 million years ago, must be ranked as one of the greatest events in the entire history of life. The theater in which it occurred certainly included present-day Gorongosa National Park. Not far away, at Malapa, South Africa, paleontologists have recently discovered a treasure trove of fossils from this critical time period. Well-preserved skeletons of a species, given the name *Australopithecus sediba*, possess features in common with other australopithecines: small total stature, chimpanzee-level brain size, shorter legs, and primitive molar teeth and heel bones. But *sediba* also has features in common with the more advanced species of *Homo:* larger overall size, longer legs, smaller teeth, projecting nose, and the beginnings of a reorganized brain. This combination of traits puts it in competition with *Homo habilis* as the direct ancestor of *Homo erectus,* and thence of its descendant, *Homo sapiens. Homo habilis* had a larger brain but lacked some of the other, more advanced traits of *Homo* displayed by *sediba.*

Experts on early human evolution have been put in a quandary by the discovery of *Australopithecus sediba.* Was this creature the direct ancestor of modern humanity, or was the true ancestor *Homo habilis*? At the least, we have new evidence of the prehuman adaptive radiation of

● ● ●

Stone Age tools can be found in Gorongosa, attesting to this area's significance in the early evolution of our species.

the australopithecines, some of the species of which were present at the birth time of the final human line.

The all-important details of the emergence of *Homo* will likely be enhanced by excavations in central Mozambique, including new sites at the border of Gorongosa National Park. The area is one of the most ecologically interesting but least explored in this part of Mozambique. It is dominated by a series of five parallel scrub-covered limestone ridges that embrace deep ravines carved into the edge of the Cheringoma Plateau and clothed in tropical rain forest. The faces of the ridges bear openings that appear to be the entrances of unexplored caves. At the terminal of one of the gorges is one such cave, from which a stream flows, and which may have been occupied by humans and the ancestors of humans since humanity began, making it a potentially prime site for future archaeological exploration.

three
─────

WAR AND REDEMPTION

Near the center of Chitengo Camp stands a ten-foot-tall, bullet-pocked slab of concrete, once part of the wall of a restaurant. It is a remnant of the 1973 attack by Frelimo insurgents during the war of independence against Portugal. Today it serves as a monument to two achievements—the birth of a free Mozambique and the rebirth of its premier nature reserve.

The restaurants and cottages were filled that day with staff and guests, the latter mostly Portuguese nationals. Suddenly a line of guerrillas emerged from the edge of the woods and began firing into the sides of the buildings, the electric lights, the windows, taking care not to hit any of the people scurrying for shelter. Their desired effect was achieved: the next day Chitengo Camp was empty. The Portuguese rulers of Mozambique, if they needed it, were further persuaded of the seriousness of the independence movement.

The military campaign against colonial rule had been launched by Frelimo (Frente de Libertação de Moçambique) on September 25, 1964. It was met by the Portuguese rulers with superior arms, scorched earth, and the

• • •

Buffaloes lying still on Gorongosa plains have been tranquilized so that blood samples can be collected for a routine health checkup.

forced settlement of rural people in fortified gulags. Portuguese secret police suppressed Frelimo activity within the cities and villages by means of imprisonment, torture, and executions. By determined incursions the Frelimo forces nevertheless progressed from their Tanzanian base into the northern provinces of Niassa and Cabo Delgado, and eventually turned the tide of battle. When the regime of the dictator Marcello Caetano was overturned by a coup in Lisbon in 1974, the Portuguese finally faced the inevitable and ceded the country to a transitional Frelimo government. The People's Republic of Mozambique was formally proclaimed on June 25, 1975.

During the 1960s, prior to the revolution, the Portuguese had made an all-out effort to accelerate the emigration of its own citizens to Mozambique. Whereas in 1940 twenty-seven thousand Portuguese nationals lived in Mozambique, by 1960 the number had swelled to over four hundred thousand. With the victory of Frelimo, 90 percent of the Portuguese abruptly left and returned to the home country. Because colonial policy had prevented the development of a black middle class, reserving virtually all jobs requiring training to Portuguese, Frelimo now faced a crisis in ordinary services. Ninety percent of the people were illiterate. There were six economists, two agronomists, no geologists, and fewer than a thousand high school graduates in all of the country. The problem was worsened when the former rulers took with them or else destroyed on the spot most of the infrastructure—vehicles, houses, machinery, and livestock—in order to prevent the revolutionary government from succeeding.

Mozambique was left bankrupt. Foreign aid could do little to improve the situation. Further, the new president, Samora Machel, was determined to change the society and jump-start the recovery by radical means to achieve national comity. "We do not recognize tribes, regions, race, or religious belief," he proclaimed. "We only recognize Mozambicans who are equally exploited and equally desirous of freedom and revolution." Brave and noble words, but unfortunately Machel determined to achieve his goals by creating Africa's first Marxist-Leninist state. Frelimo became a communist "vanguard party," and the Mozambique government signed aid agreements with Cuba and the Soviet Union. All rival political activity was outlawed. The new regime also began, in the Stalinist mode, to gather rural people into state farms and cooperatives. The policy was doomed from the outset: the rural people of Mozambique are widely dispersed in freeholds and small villages, and their entire culture was based on that traditional way of life. They could be moved only by force or inducements, neither of which the fledgling government could afford. During its brief span, the Machel plan utterly failed to achieve its stated goals. Then it was interrupted by a violent revolution.

To have a black communist state next door was anathema to Mozambique's neighbors South Africa and Rhodesia. Their white colonial rulers already faced political unrest of their own, and they dreaded the possibility of unrest turning into revolution. The global communist movement was engaged, according to South Africa's rulers, in "total assault" against their country.

The principal part of their response, exceeding even an embargo, was the creation of Renamo (Resistência Nacional Moçambicana). On paper an indigenous anticommunist army, it was in reality a well-armed military force launched by the Ian Smith colonial government in Rhodesia with the sole aim of overthrowing the Frelimo regime. When Rhodesia was liberated, in 1980, changing its name to Zimbabwe, South Africa took over as the principal Renamo sponsor. The mission of Renamo from start to finish was to achieve victory on the battlefield if possible, but failing that to destroy whatever infrastructure was left in Mozambique. At least, the reasoning went, the people of Mozambique would learn that Frelimo could not protect them.

Scars of war: the railing
of the Hippo House,
once a busy tourist
destination, buckled
under the weight of
heavy machine guns
when the building
served as a base camp
for Renamo soldiers.

It was the cruelest and most cynical of all political strategies, and it
eventually failed amid confusion and mass murder. Frelimo managed to
hold the cities and larger towns, while Renamo roamed and periodically
held and terrorized most of the rural areas. Renamo soon evolved into a
kaleidoscopic mix of coherent military units, breakaway independents,
and outright criminals. During the sixteen-year conflict, it did indeed
succeed in destroying the infrastructure of Mozambique, including almost all of its communication, travel, and trade. It left a system of feudal
states, through which both sides engaged in opportunistic attacks and
strategic withdrawals.

The total impact, to put it bluntly, was a holocaust. About a million

Mozambicans died, and several million more were forced into exile, moving to refugee camps in Malawi, Tanzania, and Frelimo-occupied parts of Mozambique. "The cruelty of Renamo mesmerized everyone in Mozambique," William Finnegan says in *A Complicated War: The Harrowing of Mozambique*, "from the peasant whose own head sank before its scythe to the members of the many foreign delegations that came to survey the wreckage." Because their purpose was essentially to destroy the country, Renamo fighters targeted health centers, schools, relief convoys, and relief workers. By 1989, more than three thousand schools had been torn down or forced to close. Innocent people were killed or mutilated for no other reason than that they were in or near Renamo targets. According to many reports, torture and ingenious methods devised for extinction went beyond ordinary barbarism to a debased, animal like frenzy. One

● ● ●

Just another school day: after the civil war ended, Mozambique invested heavily in education, and now even the most remote communities in the country offer primary education to all children.

dispatch from the United States Department of State described "shooting executions, knife/axe/bayonet killings, burning alive, beating to death, forced asphyxiation, forced starvation, forced drownings, and random shooting at civilians during attacks."

Gorongosa National Park, located well back in the interior near the center of Mozambique, was close to a headquarters of the Renamo forces, and thereby was a frequent battleground. The Renamo units at Gorongosa, who were forced to live off the land, hunted its wildlife for food. By the time the war finally came to a close, in 1992, the wildlife of the park, especially those animals over roughly ten kilograms (twenty-two pounds), had been decimated. During the political and economic crisis that followed, while the park lay unprotected, poachers and professional hunters lay waste to the remnant, with much of the meat obtained ending up for sale in the coastal cities. Philip Gourevitch, in his eloquent description of Greg Carr and the initiation of the recovery effort, describes the park as turned into fields that "shimmered with bleaching bones."[1]

Between 1972 and 2001, the number of Cape buffaloes counted in the park fell from 13,000 to just 15; the wildebeest fell from 6,400 to 1; hippos went from 3,500 to 44; and instead of 3,300 zebras there were 12. Elephant herds and lion prides were reduced by 80 to 90 percent. Of hyenas, black and white rhinos, and wild dogs, there were none.

In 2004 Greg Carr found in his first visit that he could walk or drive all day without seeing a living thing except birds. The same year, and in a contract with the Mozambican government, he committed funds, transmitted through the nonprofit Gregory C. Carr Foundation, to launch the restoration of Gorongosa National Park. When I visited the park in 2011–12, Greg was constantly on the scene as he guided the complex operations of ecological restoration. These included the expansion of the summit rain forest on Mount Gorongosa by the planting of millions of tree seedlings, as mentioned earlier, while rebuilding the tourist center to make the park financially self-sustaining. Carr provided hundreds of new jobs for villages in the surrounding region, and better homes outside the park as an option for those who were living inside. He has, so far as I have seen and heard, gained the complete respect of the people of Gorongosa and the government of Mozambique.

The megafauna of Gorongosa National Park was growing swiftly. The rate at which this was occurring varied greatly by species: elephants were at about 15 percent of their original numbers, while waterbuck had drawn to above their prewar maximum. Most wildlife species were still below half the carrying capacity. Another several decades may be needed for Gorongosa to return to its old preeminence, but given the persisting soundness of its undergirding plants and invertebrates, which survived the war intact, I believe this will surely come to pass.

My own checklist of native mammals seen while searching for ants during my five weeks' visit over two years (a naturalist without his nose on the ground looking for ants, as is my habit, and carrying powerful field glasses, would do a lot better) includes these species: porcupine, African elephant, hippopotamus, bushpig, warthog, blue wildebeest, oribi, impala, sable antelope, nyala, bushbuck, waterbuck, lion, serval cat, African civet, large-spotted genet, mongoose, bush baby, olive baboon, and vervet monkey. And human beings.

● ● ●

Somebody has to take care of
the dead, and in Gorongosa no
creature performs this task better
than white-backed vultures.

four

DUNG AND BLOOD

The near obliteration of the megafauna during the civil war and the massive poaching afterward had consequences for the remainder of the park's plants and animals that have only begun to be studied. With the zebra and antelope herds reduced to insignificance, too few grazers and browsers remained to trim the park's grasses and low-growing herbaceous vegetation. Only ground fires were left to fill that ecological role, and then mostly during the long, dry winter. Plants once kept short by constant feeding now grew higher. At the outset of the dry season, leaves and branches from bushes and trees that had fallen freely all the way to the ground were caught and held by the thickening layer of brush. The suspended detritus, kept away from ground moisture and dried by currents of air, became tinder for the lightning-struck fires. In the savannas and dry forests the flames that once hugged the ground now climbed to reach the canopies of the shrubs and trees, where, caught by the wind, they became wildfires that swept the land.

The near elimination of the elephant herds altered the dryland habitats in a different way. The big animals constantly push down middle-sized trees to bring the young and more succulent vegetation of the canopies within reach. While potentially catastrophic from the point of view of individual trees, these depredations serve the interests of the ecosystem as a whole. Thinning trees admit more sunlight to the ground and release space and nutrients to the low-growing vegetation. Wherever rainfall is low enough or seasonal enough to hold back the ever-exuberant growth of moist tropical forest, with its closed canopy and dark interior, there sprouts the great African savanna of imagination and legend.

Large animals have been bulldozing and tearing down shrubs and trees on all the vegetated continents for millions of years. Plants and other animals have not just endured the destruction, they have adapted to it. By natural selection acting on their genes, they have folded the process into the details of their physiology, behavior, and life cycles. Africa needs its elephants, or some equivalent behemoth. And so it has been around the world, where other giants have either immigrated onto the land or evolved from smaller animals already living there—giant sloths in tropical America, outsized kangaroos and marsupial "rhinoceroses" in Australia, elephant birds in Madagascar, moas in New Zealand, and, more familiarly, the array of elephants, mammoths, and mastodons that until ten millennia ago trampled the forests and savannas of Eurasia and North America.

The human assault on Gorongosa's megafauna lasted three decades. Scientists may never take the full measure of the impact on the rest of the ecosystem—on the small mammals, reptiles, and invertebrates, including insects, not targeted by the hunters. The records of their distribution and population sizes in the park, the baseline data against which comparisons at later dates might have been made, are just too sparse to allow us to draw solid conclusions. We have, however, been able to infer some of the changes that must have occurred. In an African environment well populated by big game, their droppings (often called scat) are everywhere. The size, shape, texture, and color of each tell a great deal about the animal source. A skilled tracker uses footprints, if available, to tell the species of the animal that made them, as well as its size, sex, physical condition, and speed and direction of travel. If the ground is too hard or disturbed to read footprints, the tracker turns to scats. He can get some of the same information he does from footprints, and in addition he can learn what the animal has been eating.

Those untrained in such matters, which includes almost all the rest of us, think of animal dung, if we think about any of it at all (as when scraping it off our shoes), as just a smelly mess to be avoided. But for countless small animals a scat is a treasure, a source of life. Every milligram found is a ruby, a diamond for dung feeders. From the moment it falls warm and fresh to the ground, a host of winged beetles and flies rush to feed and lay eggs on it. The scent of skatole, along with the indoles and sulfur-containing thioles, form the essence of feces. They waft downwind and repel us, but to the scatophage insects they are like the perfume of flowers. The reason for such a contradiction is Darwinian— the result of evolution by natural selection. For humans and most other animals, feces is food that we have extracted the most useful nutrients from and then discarded. It is loaded with bacteria that, if ingested, could infect and kill us. For dung-using insects, roundworms, and other invertebrates, on the other hand, feces still contain substances that can serve as nutrients. At the same time, for many of these animals fecal bacteria are part of their food.

Arriving closely behind the skatole-loving scatophages are predators and parasites that feed upon them. Most of these hunters are specialized forms of beetles, wasps, and flies. And when the scat dries and begins to disintegrate over a period of weeks, other kinds of predators and parasites arrive to replace the pioneers. As this miniature but vastly complicated drama unfolds, the inhabitants compete, fight, grow, repro-

● ● ●

A dung beetle
(*Heliocopris* sp.), with
its elytra (wing covers)
lifted, prepared to fly off.

duce, release offspring to seek other scats, then die. Simultaneously, another, microscopic world composed of bacteria and fungi churns through its own cycle, processing materials and transferring energy for the unintended benefit of larger creatures living among them.

To put the whole succinctly, the unlovely scat is an ecosystem, vital to the larger world of antelopes and lions on which we are prone to focus. It passes through a series of stages, biological and chemical, starting with fresh feces and ending in remnants of materials that sink into the soil, where it promotes the growth of plants. The history of the scat ecosystem is basically similar—on a smaller

● ● ●

The green dung beetle
(*Garreta nitens*).

● ● ●

Reaching the size of a golf ball, *Heliocopris andersoni* is one of the largest and most impressive dung beetles in Africa.

scale of space and time, of course—to the transformation of a pond into a bog, or to a pasture that turns first into a grassy meadow and then into a forest.

A parallel class of ecosystems occurs in the decay of bodies. In a healthy African savanna environment, most of the remains of large animals are quickly consumed by predators and scavengers, principally lions, leopards, wild dogs, jackals, and vultures. The sparse rotting fragments, consisting mostly of bone marrow, fascia, and cartilage, are attacked by a succession of small organisms, which are different from

• • •

More than six thousand strong, the waterbuck population has nearly doubled compared to its prewar size. If not for the ecosystem services provided by dung beetles, the grasslands of Gorongosa would quickly drown in the waste produced by these efficient grazers.

those in scats.

When the Gorongosa megafauna was almost destroyed, the scat and corpse decomposers were comparably diminished. It might seem likely that this segment of the fauna suffered many species extinctions, because they are a link in the food chain below the megafauna and dependent on the widespread distribution of products provided by the big animals. It is further true that scat and carrion feeders must locate them at distances proportionate to their small size, which are vastly greater than the distances facing the large animals that provide the resources. A blowfly, for example, traveling one hundred meters to reach its odoriferous goal must, when corrected for body length, cover the equivalent of about twenty kilometers (roughly twelve and a half miles)—in other words, from one end of the park halfway to the other.

Other flying insects, mostly flies and beetles, solved the problem brilliantly in an early stage of their genetic evolution. In the case of beetles the achievement was probably made as far back as the Paleozoic era, more than 300 million years ago. When the adults detect the odor of decomposition, they take wing and travel upwind. Even a faint breeze bearing an infinitesimal scent can bring them quickly to the source. At long distances, it might seem the insects need a gradient, allowing them to orient from a lower intensity of the odorous molecules to a higher one, then still higher, until they arrive at the source. But when molecules are dispersing by wind in such low densities over long distances, this reading of a gradient may be impracticable. So the insects have evolved a simpler and more powerful method: fly upwind as long as you detect the smell of the target. If you lose the odor, no problem. Just fly side to side until the smell is detected again, then continue upwind. By zigzagging, dung and carrion lovers are able to arrive at the target with remarkable speed, even when, in some cases, they must traverse straight-line distances thousands of times greater than their own body lengths. The method works even in a shifting wind. In that case the insects find the odor stream simply by widening their search pattern.

You may have noticed that upon entering a meadow, woodland, or even your backyard, which at first glance seems free of insects, there will soon appear a great many of some bloodsucking fly or other, whether deerflies, mosquitoes, buffalo gnats, or ceratopogonid midges ("no-see-ums"). They are likely to arrive in varied combinations together, or else sequentially around the clock. My own personal record for such airborne assault occurred one late June afternoon in a meadow on the edge of forest in the Upper Peninsula of Michigan. During a three-hour period, in a nice-looking spot where I and two other naturalists had chosen to camp, buffalo gnats, deerflies, mosquitoes, ceratopogonid midges, and snipe flies belonging to the taxonomic family Rhagionidae crowded in to find a space on uncovered parts of our skins. We lost some blood, but at least in exchange we obtained some of the rarely seen snipe flies for Harvard's Museum of Comparative Zoology.

Never during all my travels in tropical forests and swamps have I experienced the sanguinary equivalent of that onslaught in Michigan. Yet even there, while the bloodsuckers so earnestly sought our company, I was aware that they represented only a tiny fraction of the huge di-

Egyptian mosquito (*Aedes aegypti*).

versity of insects living in the same place. They were greatly outnumbered, for example, by ants, grasshoppers, aphids, springtails, and other kinds of insects that swarmed through the meadow and forest.

How was it possible, then, for the bloodsuckers to find us so quickly? The reason is that they use the same rule of thumb as do the insect seekers of feces and carrion. Human beings emit streams of more than one hundred kinds of airborne odorants, from carbon dioxide to complex hydrocarbons. These inadvertently broadcast our presence to the insects that depend on animal blood and have accordingly evolved a hyperacute sense of smell—our smell.

In the dry forests and savannas of sub-Saharan Africa, the first such specialist you may feel landing on your neck or arm is frequently a tsetse fly. When I first slapped one and examined it in the palm of my hand, I thought, So this is the great scourge of Africa. The twenty-three species of tsetse flies known are each about the size of a housefly, but easily distinguished by their long, biting proboscis, which extends straight forward from the head like the needle of a dart.

Tsetse flies are, paradoxically, the principal reason why wildlife sanctuaries like Gorongosa exist in the first place. Early European settlers found that the grasslands and savannas of Africa are ideal country for cattle, just as they are ideal habitat for antelopes, Cape buffaloes, and other large native herbivorous mammals. The African drylands, however, are also home to tsetse flies—and to the trypanosome protozoan they transmit that causes sleeping sickness, lethal to both cattle (in which it is called bovine trypanosomiasis) and human beings. The symptoms are anemia, lethargy, and progressive and irreversible weight loss. Indigenous African mammals, including oft-bitten antelopes and Cape buffaloes, tolerate the infection, but most strains of domestic cattle cannot. As an unintended result, people today can go to the African parks and experience Earth's Pleistocene environment in almost all its glory.

Like creatures dependent on scat and carcasses, tsetse flies and other bloodsucking insects must have been diminished to some unmeasurable degree by the near erasure of the Gorongosan megafauna. However, the remainder of the park's invertebrates, including other kinds of insects, probably remained intact. The reason is that the vegetation and soil of the park were only slightly altered by the civil war. Smaller creatures, from spiders and ants down to nematodes, fungi, and bacteria, the "little things that run the world," as I like to call them, maintained the ancient biochemical processes that drive the ecological cycles. The same stability favored life that continued to teem in the rivers and lakes. When I first visited Gorongosa, in 2011, the park was ready and able to receive the returning megafauna, along with clipped grass, broken trees, scat on the ground, bloodsuckers on the wing, and all the rest of the foundation of everything so enduring and magnificent.

● ● ●

Tsetse fly (*Glossina* sp.).

●●●

Nile crocodile.

five

THE TWENTY-FOOT CROCODILE

One species of the megafauna, the Nile crocodile, withstood the slaughter of big animals occurring all around it during the 1976–92 civil war in Mozambique. The big reptiles are a good potential source of meat and hide, but not so easy to hunt, bring down, and retrieve. On the land, which means the banks of rivers and lakes, they are timid. At the approach of humans on foot or by helicopter, they slide into the murky water and swim away from the shore. All that can be seen then are eyes, nose, a strip of back, and a gently swirling tail. They can be killed with a precise rifle shot, but the hunter is then faced with the serious problem of how to retrieve the body when other big, live crocodiles, unafraid in their watery stronghold, might be lurking nearby.

So the crocodiles of Gorongosa have been permitted to reach their full life span, which averages forty-five years. Many among the clusters gathered along the waterways are both old and very large. One I saw traveling overland was "One-eye," a snaggle-toothed female that looked well fed despite her visual impairment. Exactly how large crocodiles get was a matter of speculation while I was at Gorongosa. Fifteen feet and five hundred pounds might be the maximum, but nineteen feet is possible. Larger than that is only remotely possible; twenty-foot Nile crocodiles have been claimed elsewhere but remain an unproved legend. The claimed limit to size in this species of crocodile is curiously similar to that in the great white shark. The usual maximum length of the great marine predators is around sixteen feet, but twenty feet or longer is remotely possible. Stories of still larger size have been told, with a few monsters said to exceed thirty feet, but experts dismiss the claims as either mistaken or outright fictitious. Of course, a swimmer or capsized fisherman would not want to encounter a Nile crocodile or great white shark even if it were only a relatively juvenile ten feet. Meanwhile, I hope that measurements can be made of the animals resting on the banks of the Mussicadzi, Urema, and Púnguè rivers. To avoid the risk inherent in using a tape measure on foot, it might be possible to get the job done by triangulation from a helicopter. Maybe there is a twenty-footer somewhere in Gorongosa. I hope I will get to see one.

Crocodiles and hippopotamuses congregate close together in the waterways of Gorongosa. The adults of both species are formidable giants, each capable of mortal damage to the other. A large crocodile can take down a lion, if one can be approached while it drinks at the water's edge. A big hippo can kill almost anything, with huge tusks driven by sledgehammer swings of its immense head. Fortunately for all, such fights seldom occur.

Perhaps the almost intimate coexistence of the two giants illustrates a principle of evolutionary biology: size counts. One stratagem to defeat predators is to rise, literally, above your enemies. A third member of the (almost) impervious titans are the elephants, which drink and bathe, slowly and casually, among the crocodiles and hippopotamuses. The corollary of the adaptation is that if you are not big enough, evolve to run fast and be prepared to fight ferociously if cornered. That is the Cape buffalo and human way.

A principle of ecology, the stable materials cycle, is also illustrated by the harmony of the giants. At Gorongosa the cycle connects five elements, hippo to plant to fish to crocodile, then finally to large birds and back to hippo. The hippopotamuses are nocturnal grazers. They come onto the land at night and waddle over to "hippo meadows," as much as a mile distant from the water, where they feed on grasses and herbage kept clipped low and productive by their activity. They consume only about 1 percent of their weight during each foray, but that is still a lot of forage when eaten by a dozen or so animals in a pod, each weighing

● ● ●

A young conservationist, Tonga Torcida, and a sign warning villagers about the danger lurking in the river—each year several people in Mozambique fall victim to crocodiles.

from one to somewhat over two tons. Through the day the hippos stay in the water, defecating copiously, and thereby nourishing algae and aquatic plants. These organisms support teeming populations of aquatic insects and herbivorous fishes. Among the latter, the waterways harbor species of catfish, barbs, suckermouths, mullet, and tilapias. These in turn support predators such as bream and tiger fish. Also among the predators are the crocodiles. The big reptiles, however, are not entirely dependent on fish. They are highly generalized carnivores, feeding on any kind of animal they can catch and kill. Among their occasional prey are people who get careless while fishing or washing laundry in crocodile habitats. In the Gorongosa area, the big reptiles account for a majority of the mutilations and deaths caused by wild animals as a whole. For their part, these top-level predators serve the equally top-level herbivore hippos by limiting the influx of other grazing animals onto the prime meadows. Crocodiles also help to serve the plants and fish by the feces they release into the water. Finally, crocodiles are not the exclusive top predators. As newly hatched babies they are food for large water birds such as the marabou stork and goliath heron.

• • •

Filmmaker Bob Poole holding all that remains of this blind, from which he intended to film crocodiles, after one of the animals dragged it underwater and devoured it.

Did the near extermination of the hippopotamus by soldiers and poachers break the cycle? No, it just became four-sided instead of five-sided for a while. And was the biodiversity of the aquatic and shoreline ecosystems perturbed in any lasting manner? Probably not, because the absence of hippos was relatively brief. But we will never know for sure, because the necessary prewar baseline research was not conducted. In any case, the aquatic ecosystem has now been restored. And now that the rain forest of Mount Gorongosa is also saved, and providing the water distribution of the park is protected, and providing the long-term effects of climate change are not overwhelming, there is no reason to fear that the ecological cycle that worked for thousands of years in the past will not go on for thousands of years into the future.

• • •

Joyce and Bob Poole.
Photo by Sarah Arnoff.

six

THE ELEPHANT WHISPERER

J oyce Poole is one of the world's leading authorities on the African elephant. Her interests run deep into all aspects of the species' history, social behavior, and family life. Because she is further concerned with fine details of the elephant-human relationship, she could be called an elephant psychologist. I prefer to think of her as the elephant whisperer.

Her services were much needed in the rebirth of Gorongosa. When the park was in its prime, with dense populations and a diversity of large animals that was among the highest in Africa, elephants were dominant animals. They roamed the land in herds up to sixty or seventy individuals, each loosely organized into clans of mothers with offspring, their sisters, grandchildren, and often close collateral relatives. All this came to an end during the civil war and the period of chaos immediately following. Soldiers and poachers hunted down and shot a great majority of the population in the park.

Elephants are highly intelligent animals, and relationships among the members of the clans are intimate and long remembered. Each clan consists largely of females and is led by a matriarch, usually the eldest female. Young males leave the clan in early maturity to form groups of

•••
Gorongosa elephants. *Photo by Jean-Paul Vermeulen.*

their own, sometimes returning to revisit their birth clan. During the season of hormone-driven musth, adult males are admitted to the clans in order to mate.

By 2011, during my first visit to Gorongosa National Park, the populations of most of the surviving mammal species had reached about 10 percent or more of their prewar levels. It is a safe bet that few if any of the animals of almost all other species retained memories of what happened to their kind by human depredation. But that was not the case for the elephants. With natural life spans of half a century or longer, and legendary memories, the survivors remembered the horrors inflicted by humans on foot and from motor vehicles.

Among elephants still alive as the restoration of the park got under way, a few were old enough to be either matriarchs or lone wandering males. As youngsters they suffered what appears to be the equivalent of post-traumatic stress syndrome much the same as that experienced by human war veterans. These elephants, as Joyce Poole puts it, "have seen terrible things." The aging matriarchs of Gorongosa lead clans that in Africa are among the most skittish and likely to charge if they feel threatened.

The solution to the problem, according to Joyce Poole, is to teach the elders, and through them the younger members of their clan, that "not all people are bad. We can be trusted." The method is basically through what experts on animal behavior call habituation—the calming effect of repeated peaceful contact. It is the reason that snake charmers and some religious cultists can handle snakes safely—most of the time.

To achieve this result, Joyce Poole tries to get into the elephants' heads. She and her photographer-naturalist brother, Bob Poole, sit in their special elephant-resistant vehicle, roll-barred and frame-welded, as close as possible to the herds, doing nothing most of the time, just sitting and talking. They walk about casually in the manner of, say, peace-loving marabou storks or waterbucks. They adjust their approach to what they can see at a distance of the elephants' mood, estimating the best approach and closeness they can safely take, one day at a time.

By and large this elementary yet sophisticated method has worked. The herds are more approachable, the matriarchs less excitable. None-theless, the sight of humans and automobiles still evokes occasional sus-picion and hostility: trunks raised into the "periscope" position to sniff incoming odors at a distance, approaching closely with ears flared, and "staring down the trunk" at intruders that don't retreat.

For a long time Joyce and Bob Poole thought that attacks by el-ephants in Gorongosa were a thing of the past. They and others who came close to herds with anything less than extreme caution still received occasional threats of a magnitude to be expected of elephants anywhere. Once an aging six-ton male ran parallel to the Poole vehicle in the man-ner used by bulls to intimidate rivals during musth. The behavior was not play. It was a serious threat and a close call. An all-out attack by an animal of this size might have destroyed the elephant-resistant vehicle, with possibly fatal results.

In a second incident a large female, probably a matriarch, ap-proached the vehicle to within a few feet, ears spread, and stamped her feet. It was frightening but short-lived, because the big animal soon turned back.

● ● ●

Decimated during the war, the Gorongosa elephant population is growing rapidly and is now approaching five hundred individuals.

Then, finally, the worst happened. Bob was returning one night from the field with Greg Carr. Also aboard were other park managers and a guard armed with a gun. The car was stopped by the trunk of a tree pulled down across the road by a feeding elephant. Without further thought, Poole took a detour by steering the car into the bush. Suddenly three elephants close by began screaming. One, perhaps the matriarch, ran up to within several feet of the car, put her head down, and rammed the vehicle on the left side at the level of the engine. The blow smashed the left fender and protective roll bar and blew the tire off the rim. Bob started to back up to give the elephant room and incentive to pull away. But the enraged animal then attacked the front of the car, folding back the defensive bars and bringing her head to within inches of Bob, who

• • •

These juvenile elephants demonstrate that the park's large animals are returning. *Photo by Joyce Poole.*

was now pressed against his seat. At that point, following procedure, the guard fired a shot over the elephant's head. Paying no attention, the elephant began to push the vehicle backward, with the people inside now completely helpless. Her charge ceased only when the vehicle struck a tree, at which point the guard fired a second warning shot. (The official procedure is three warning shots, and only after these a shot to kill.) The elephant, unable to push farther, backed off and walked back into the dense forest.

What caused this rare and potentially fatal charge after years of a perfect record? One factor was likely the accident of surprise to the elephants as the vehicle wheeled off the road and into the forest toward them. Another might have been the smell of the gun itself. Even a slight whiff as the car approached might have stirred old memories. In any case, Joyce and Bob Poole hope that the truce they have helped forge will now persist as younger elephant matriarchs take control of the herds.

•••

The skeletal remains of the Hippo House, once a busy restaurant and observation point.

seven

THE HOUSE OF SPIDERS

At the end of a long rutted road in the park sits a conspicuous artifact in the midst of wilderness. Built in 1970, the Hippo House was the vantage point, the *antigo miradouro*, from which well-heeled tourists, cool drinks in hand, watched wildlife herds as they grazed over the vast floodplain grassland below. Today the herds are back, but the house is a seldom-visited ruin. During the Mozambique civil war, almost all the buildings of Gorongosa National Park were torn down or blown away, leaving behind a few remnants scarred by bullets. The house had been reduced to a shell of its original self. Of the two floors, only the concrete upper one and the corner posts supporting it survived. The lower floor, while still sheltered by the ceiling above it, was open on all sides.

When I first visited the Hippo House, Mozambique was in the middle of the winter dry season. Other than along the watercourses, the vegetation of Gorongosa was brown and withdrawn. Insect life was still abundant, but harder to locate. I had been told that spiders, big ones, were abundant at the house, but I was quite unprepared for what I found. The interior of the ruined building was powder dry. Its floor, stanchions, and ceiling were windblown and coated with dust. No vegetation reached in from the outside, and except for a few small geckos resting on the pillars, there was no immediate sign of life of any kind. Instead, torn webs and long single threads of silk dangled from the ceiling like ghostly decorations in a haunted house. They swung gently back and forth in the occasional light breeze. No other movement or sound came from the seemingly empty space.

Where were the spiders I expected? Not one could be seen. But I knew they must be there someplace, alive, perhaps watching us. The idea of a hidden arachnid horde ready to rush out made me uneasy. Soon I saw something else: round objects plastered onto the ceiling. They were dusty and silent. My companions and I picked up a stick lying on the ground outside that was long enough to reach the ceiling, and tore two of the pouches apart. They proved to be silken egg pouches, undoubtedly made by spiders but now dry and empty; we were obviously not in the breeding season. The spiders themselves stayed hidden. Where were they? I grew more apprehensive.

We saw other, much larger, oblong pouches scattered over the rough eroded ceiling. At the tip of each was a circular entrance opening to a hollow interior. Using a flashlight and looking straight down the chamber, we could see what lay within. There at the rear of each pouch crouched a large spider, facing outward, its fangs, eyes, and the front of its tightly bunched legs visible. I wanted to see a specimen well enough to identify it, but hesitated. I was, to be frank, afraid of these crouched and waiting spiders. I suffer from mild arachnophobia. This spooky place was the setting of an arachnophobe's nightmare.

We selected one of the pouches and poked at it in and out, but the spider stayed tight inside. One of my companions then took charge. He tore open the pouch and shook the inhabitant out into a transparent plastic bag. At last I could see what had lain within. The spider was heavy-bodied, the size of a thimble. When it suddenly spread its spiny legs, its width almost tripled.

Our captive soon lay quietly in the bag. I would have preferred it to scramble around in a frenzied panic, as most spiders do, but it just stayed ominously quiet, as though composed, and waiting. Waiting for what? I thought. I had to admit that it was pretty in its ghastly way. The top surface of its abdomen was a bright tessellation of white, yellow, and brown patches. The pattern they formed camouflaged the spider against the dull, grainy surface of the house ceiling. In the center of the lower face of the spider's abdomen was a bright red spot. I said to myself, Yes, of course, just like a deadly black widow. The colors presumably give a warning to possible predators when the spider is out of the pouch: stay away; you touch me at your peril.

I had solved the mystery of the spider house, at least in theory. The creatures in the silken bags were orb weavers, members of the spider family Nephilidae, called golden orb weavers, and, I later learned from an arachnologist, the species is *Nephilengys cruentata*. Some species of nephilids and the closely related araneids hide in retreats next to their webs; others remain in the centers of the webs. The Hippo House spiders stay in their retreats until nightfall, and come out at night, when birds and other spider-eating predators cannot see them. Then, like orb weavers everywhere, they spin their webs. When a moth or other flying insect blunders into the web, the spider wraps it in silk sheets, carries the immobilized prey back into the retreat, and eats it.

• • •
Orb weaver
(*Nephilengys cruentata*)
from the Hippo House.

I surmised that the big females like the one I now held in my hand accept much smaller males during the rainy season, when savanna forests and grasslands are green and insects fill the air. After mating, they spin the round egg sacks we found and deposit their egg masses inside.

•••

An unlucky katydid that flew into a web under the Hippo House is immediately killed and wrapped in silk by a female orb weaver.

But how could there be so many spiders of this one species crowded together? Why are there no other creatures of any kind? The explanation I believed to be immediately clear. The floor of the lower level of the spider house is a layer of concrete. The interior is abnormally dry. Because the lower level cannot be invaded by any vegetation, few if any other forms of insect or arachnid life can live there. Yet flying insects undoubtedly fly through the wide-open space of the lower level, in through one side and out the other. A few might settle there to rest. The fate of most or all is the same: spider food.

My imagination was roused by this bizarre little world, but more so by my own reaction to it. When I took the captured spider back to Chit engo Camp, I found I was unable to make a specimen of it. That would mean fishing the monster out of the cellophane bag and working it into a bottle of preservative. So I simply opened the rear window of my room and dumped my captive live onto the ground below, where it would at least have a chance of making its way to a tree or building and spinning a new silken retreat.

I remember vividly the incident that made me an arachnophobe. I was eight years old. It was late summer, and I was exploring a vacant lot near our house. There were several full-grown female orb-weaving spiders in the high weeds, likely the common garden spider (*Araneus diadematus*), sitting in the center of their webs. To an adult these biggest of American spiders, at least biggest next to tarantulas and fisher spiders, are intimidating, but to an eight-year old they were terrifying. Although only about two inches long, excluding outstretched legs, each seemed the size of a human hand, and their webs seemed the size of a door. I could not resist getting close enough to see all the details of one spider's body. The giant had been sitting quietly. When I was about a foot away, it began to jerk back and forth in a menacing manner. I thought it was preparing to jump out and onto me. I ran. If that were not bad enough, I soon afterward saw a movie, the name of which I have long forgotten, in which a man is trapped in a cave. Blundering around, he becomes tangled in spiderwebs that are hung all around. Spiders, really big ones, climb down toward him, and . . .

●●●

Tarantulas, known in southern
Africa as baboon spiders, may
look frightening but are generally
harmless. Their main line of
defense is not their venom, but
tiny urticating hairs that cover
the entire body.

True enough, arancid spiders like the ones in the weed lot and in Gorongosa bite only if handled, and they are not poisonous, either. But this knowledge, which I possessed at the spider house, did little to blunt my lifetime aversion. That deep, autonomous response has done some good. It has made me more careful with black widows, brown recluses, and the small minority of spider species whose bites can do serious damage.

Aversions and phobias of this kind, with the latter an extreme response causing panic and cold sweats, can be imprinted with as little as a single brief episode. They are rarely caused by a frightening experience with a knife, a gun, an automobile, or any other modern contrivance that can injure or kill. On the other hand, they easily and quickly follow a frightening experience with one of mankind's ancient perils: snakes, spiders, wolves, heights, running water, and closed spaces. Psychologists have determined that the most powerful stimulus in establishing a snake phobia is the overall appearance of the animal. Otherwise there would be a constant plague of eel and earthworm phobias. The key experience is to be startled at an early age by something moving, looped, or coiled, close by on the ground, especially if the child's companions express fear.

The ease with which an individual, especially a child, acquires such a strong aversion or full-blown phobia varies a great deal among people, and at least part of the variation of the susceptibility is due to personal hereditary differences. It is reasonable to interpret this pattern of selective phobias—spiders but not ants or butterflies, and snakes but not eels—to be the consequence of evolution by natural selection. During millions of years of human prehistory, it has paid in a major Darwinian way to have quick, decisive response to the things that can kill you.

So I have forgiven myself for the wavelet of fear and revulsion I felt about the harmless denizens of the spider house. Let me make further amends by stressing that people are mostly safe amid what remains of living nature. We conquered the man-eaters long ago by destroying almost all of the big predators willing and able to hunt humans. They survive in our stories and in our legends of monsters. We imagine them silently emerging from caves and swamps, easing up from unexplored depths of the sea, or drifting down unseen from above.

But the fact is that the only true man-eaters left are tigers in the swamps of the Sundarbans; the Nile and saltwater crocodiles; and three species of big sharks, in order of importance the great white, tiger, and bull sharks. These are rare; except for the crocodiles it takes a special season even to get a glimpse of one. It is a disappointment for me to see a large crocodile slip away and hide in the water as I approach, even at a distance. I want a better look. In Australia, where saltwater crocodiles are famously dangerous, attacks cause an average of only one death every two years. In Africa, the fatality rate is much higher, into the hundreds, but still a relatively small source of mortality. Also, about the only sure way ever to glimpse a great white shark is by guided tour off the coast of South Africa. The total number of people killed by all kinds of sharks each year worldwide has varied between four and eleven. The event is so rare that each one usually receives worldwide press coverage. Tigers in the wild, despite their rarity, account for more than a hundred fatalities a year, mostly in India; lions fewer than that; and free-ranging leopards and jaguars still fewer. More people are murdered in any American big city in a year than are killed by the tiny remnants of big predators still alive.

That said, I regret I cannot give a similar pass to venomous snakes. I spent a good part of my teenage years hunting snakes in the woods and swamps of Alabama and northern Florida. I discovered something any

•••

The golden orb-weaver (*Nephila senegalensis*) is one of the largest spiders of Gorongosa. Its name comes from the beautifully golden coloration of its silk.

herpetologist can tell you: unless you know where to look for particular species, which streamside to wade along or which log to roll over, and what time of the day or night, you will see very few snakes during a woodland walk. The same is true in most parts of the tropics. Unless you are a practiced snake hunter, it is a treat just to see a snake of any kind on a given day.

Unfortunately, some of the snake species on all the continents (except Antarctica, of course) are venomous, and while it might be a rare day that even a veteran woodsman accidentally steps on a large one, the risk is there. Germany, France, and England each record about a dozen fatalities a year from snakebite on average, and even unlikely places like Sweden, Finland, and Mongolia record one or two. The highest risk is in tropical Asia, with more than ten thousand deaths from snakebite each in India and Indonesia, and in sub-Saharan Africa, led by ten thousand fatalities in Nigeria. Partly responsible for these tragedies is the close contact of large populations with snakes in rural areas. Oddly, despite Australia's deserved reputation for large, belligerent venomous snakes, fewer than a half-dozen people die from their bite each year. Exact figures from all countries are impossible to come by, but average yearly snakebite fatalities worldwide could be as high as ninety thousand.

Walk into or swim in any wild habitat remaining on Earth, maintain the same level of caution you would on a city street, and you will be far safer than in most urban environments. Use common sense: don't swim with crocodiles; don't paddleboard among seals where great white sharks have been seen; and above all, never, ever run up to a mother grizzly bear with cubs to take a better look. Your greatest risk in the wild is from insect-borne disease—malaria, dengue, leishmaniasis, yellow fever—and these can be deadly if untreated. But they are transmitted chiefly among people. They can be easily avoided, and in any case pose less risk to you than the mélange of pathogens passing directly from person to person in human settlements.

The scary but harmless spiders in the Hippo House, and all the other animal species of wild environments like those of Gorongosa, are instinct-guided. They rigidly follow life-and-death routines formed during millions of years of evolution. Their lives are finely tuned and fragile in ways that are blessedly unthreatening to human beings.

eight

THE CLASH OF INSECT CIVILIZATIONS

Anywhere I am in the world I love it when the air is warm and moist and heat bounces off the sunlit earth, and insects swarm in the air and alight on flowers and come in droves to lights at night. I remind myself that their species number is in the hundreds or thousands. None to me is a bug. Each instead is one kind of insect, the legatee of an ancient history adapted to the natural world in its own special way. I wish I had a hundred lifetimes to study them all.

Social insects, flourishing at the peak of their 350-million-year history, are the most spectacular in all the great faunas on the land. Their species variously build cities, raise gardens, and conduct endless wars to acquire territory or capture slaves. Their civilizations are built entirely with instincts that guide them through their epic lives. The wars they conduct are worthy of Homer.

They change the landscape. The scrub savanna and dry deciduous

● ● ●

At the risk of painful stings Tonga Torcida watches Matabele ants returning from a successful raid of a termite colony.

forest of Gorongosa National Park are peppered with cone-shaped hills ranging in size from a refrigerator to a city bus. The largest is about ten feet high and thirty feet or more across. From most grow a variety of plants, including shrubs and trees. Birds rest on them, and animals climb up to the top to scan the surrounding terrain. But the real action is inside. There, in a labyrinth of rooms and corridors, live up to a million termites of the genus *Macrotermes*.

The macrotermitine termites are the sole architects of these mounds. They have changed the mostly flat surface of the Rift Valley and surrounding plateaus into a land of varied contour. They provide an environment that enriches the fauna and flora. The density of the mounds at Gorongosa ranges from one, or sometimes none, to over fifty per hectare (about two and a half acres). Each of the bigger ones is a hillock to human perception

● ● ●

Digging for entomological treasures, the author uncovers a colony of fungus-growing termites.

but to the termites a mountain, more precisely an inselberg, or small, isolated mountain. Each mound nest rises as the dwelling of a single colony. When the colony dies, the nest slowly erodes and the earthen material from which it was built very slowly dissipates back into the soil. *Macrotermes* colonies do not meet and parlay, nor do they exchange workers or occupy abandoned nests. Their territorial boundaries are absolute. For *Macrotermes* the nest is the colony, and the colony is the nest.

The living colony itself is a tightly organized superorganism. In a large royal chamber at the center of the nest sits the mother queen. From her virgin youth, already big for a termite, she has grown to the size of a human thumb by the swelling of the ovaries in her abdomen (the rear major segment of the body). Her abdominal volume reaches ten milliliters (about one third of an ounce), making her the largest of all social insects. Surprisingly, these obese giants are still able to move very slowly under their own power by a combination of leg traction and forward-moving waves of peristaltic contractions of the abdominal muscles. Each queen is accompanied by her much smaller mate. The royal pair is fed continuously with salivary secretions by a large retinue of nurse workers crowding around them. A queen in her prime has one of the highest reproductive rates ever recorded in land-dwelling animals—according to one estimate, up to thirty thousand tiny eggs a day, or ten million a year. The output continues for much of the life span of up to twenty years. This high output is needed to create and maintain the population of the all-neuter workers, which compose the entire workforce.

The nest is also a garden. All through the core, a foot or more in from the outer earthen crust, are located chambers filled with spongelike masses of partially digested pellets made in turn from dead wood, grass, seeds, and other diffuse sources of cellulose. Growing on the masses are symbiotic fungi of the genus *Termitomyces*. The termites eat the masses, fungus and all.

• • •

A fungal garden found
in an underground
termite colony.

The termite colony is a factory within a fortress. It is defended from its many enemies by an army of soldiers, equipped with hard cylindrical heads and sharp pincer jaws. More abundant are the round-headed workers, responsible for round-the-clock labor. A large force of male workers forages at night for bits of dead leaves and other woody detritus. Inside the nest this material is eaten by the inhabitants and is also fed to the symbiotic fungus. Female workers, slightly smaller in size than the males, attend to the domestic chores, including the construction of the nest walls. The material used for construction consists of saliva and excrement mixed with soil. When a hole is created anywhere in the nest walls, say by a falling tree branch, a termite-eating aardvark, or an inquisitive entomologist with an adze, soldiers rush to it and mill about, ready to attack the intruder. Soon masses of workers appear and start to repair the cavity by laying pellets of fecal matter along its edges. If the breach in the crust is not very large, say no more than two spread hands, it will be sealed within several hours.

The nest is an air-conditioned building. It is constructed so as to produce a constant flow of freshened air through the living quarters and fungus gardens of the nest core. As air is warmed in the core by the metabolism of the huge colony and masses of fungus, it rises by convection to a large chamber in the upper part of the mound, then passes out to a flat, capillary-like network of chambers next to the outer nest wall. Here the flowing air is cooled by radiation to the outside. It is also refreshed, with excess carbon dioxide diffusing to the outside and oxygen diffusing to the inside of the nest. As its temperature drops, the air sinks and flows to large lower passages of the nest beneath the central core. Through this upstream the temperature in the living quarters and fungus gardens is kept close to 30°C (86°F) and the concentration of carbon dioxide is held at about 1.3 percent.

The mound-building termites are not instructed by architects, nor do they work off a blueprint in their tiny brains. So how do they build these elaborate structures that so dominate much of the African landscape? The key was found in an experiment performed by the great French entomologist Pierre-Paul Grassé in the 1950s. He began by removing workers from their mound nests and placing them in a container along with piles of pellets of building material consisting of excrement and soil. At first the termites wandered about, obviously disoriented. Then individuals began to pick up pellets, walk about some more for a short while, and finally put their burden down haphazardly on one open space or another. Their actions watched by Grassé were at first not coordinated. A pellet deposited by one worker was likely to be picked up by another worker, carried about, and put down somewhere else, to be picked up by yet a third worker. Finally, by chance, two or three pellets got stuck on top of one another. The little pile proved much more attractive to the termites than single pellets. The workers began to add more pellets on top, and a column began to grow. If two columns were constructed close to each other, the workers on top of each column were attracted to the other, so that after a certain height was reached, the workers began to place the pellets in a way that bent the tops of the columns toward each other. Soon the growing ends met, creating an arch, whereupon the builders moved away to work on other columns and other arches. Because termites are blind, they most likely were orienting by the attractive odor of the pellets—although some amount of signaling by sound was not eliminated.

An array of arches is still far from a completed nest, but the manner in which these basic structures were put together provides a clue to how such complex architecture is achieved by tiny brains. It is genetically programmed self-organization, the same principle by which cells are built from molecules and organisms are built from cells. The termite algorithm is as follows: At the beginning, while there is little or no order, the termite takes the first step. When that is completed, it takes the second step, and so on, one step at a time, each triggered by completion of the first and receipt of stimuli created or amplified to that point. The algorithm commands: Proceed to the next step, and at the end of the sequence, stop. You have a nest. No termite holds a picture of the finished product in its brain. When activated, the termite proceeds through a series of simple actions. At the end of the sequence it halts and begins some other routine. When many termites proceed together, even so complex a structure as a mound nest can be quickly built.

The mound-building termite colony is a superorganism at its most elaborate. Like an organism, it has a birth, an adolescence, and a maturity. At maturity it begins to reproduce, releasing virgin queens and kings to start new nests. After a preordained period of time, it dies. The death of the colony follows upon the death of the queen, who lives for a maximum of fifteen to twenty years. With the queen gone, no more sterile soldiers or workers are produced. Since they have a much shorter life span than their mother, the nest population soon spirals down, and when the last inhabitants die, they leave a lifeless mound behind. During the years to follow, the mound is slowly eroded down to the ground level, continuously releasing its valuable nutrients to the soil. At any given time the mounds strewn across the savanna and forests are very roughly half living and half dead.

Each colony, each closed population of mound builders, begins with a queen and her much smaller consort king. Newly bonded, they search on the ground together for a suitable spot and, upon success, dig a small vertical tunnel. Of all the virgin queens and males who depart from their mother nests to start a new colony, only one pair in many thousands reaches even this earliest stage. Everywhere the pairs go, to mate and search for nest sites, they are decimated by predators—birds, lizards, centipedes, insects of many kinds, and, not least, humans. People and their prehuman ancestors have probably feasted on royal termites for hundreds of thousands of years. Even after colonies have increased in size and managed to build a first, golf-ball-sized nest, most soon perish at the hands of enemies skilled at digging them out. Once the nest grows large enough, however, and the soldier force is strong enough to repel invaders, the mortality rate of colonies drops sharply. From the start the royal pair and their neutered minions have been in a race to achieve bigness. Exponential growth is the key to survival. Within four to six years, the upper population limit is reached at one million workers, possibly more. At this point, the colony starts to invest heavily in the production of winged virgin queens and males. It prepares to reproduce itself. Then, on certain days, at a particular hour, under the right weather conditions, the royals emerge in vast numbers to fly upward and away, catching wind currents, traveling for a distance of up to several kilometers. Then each lands, and each of the extremely lucky few to survive finds a mate and begins its own epic.

For millions of years, long before the australopith ancestors of humanity walked among their mounds, the *Macrotermes* termites dominated the drylands of Africa. The fauna and flora adapted to them, came to depend on them to scavenge land vegetation, used their mounds to get above ground level, and flourished in the special soils they created. A few animal species specialized in evolution to seize and feed on some part of the ten-kilogram (twenty-two-pound) mass of termite tissue in each colony.

The termite mound, like so many segments and products of its living environment, is a small ecosystem in itself that progresses through a programmed cycle. As it rises from flatland to towering hillock and recedes back to flatland, it creates a sequence of microhabitats. As the little mountain rises, early erosion of its lower wall forms an encircling pediment apron with a different chemical composition from that in the soil just beyond. The apron generates its own miniature ecosystem, while high above, the summit of the mound attracts birds, monkeys, and smaller mammals, which look around from it in relative safety. These visitors add nutrients in their waste material, and drop seeds able to germinate variously on the mound and around its perimeter. The vegetation that sprouts from the hill consists mostly of shrubs and little trees, which attract still more birds and mammals, along with a wide variety of insects and other invertebrates. Often there is a single larger tree near the center that has taken root during the early development of the mound. The hill-borne vegetation attracts antelopes and other browsing mammals. In subsequent years the condition of the termite mound begins to decline. Overbrowsing by the large herbivores opens the vegetation and wears down and cracks the mound surface. By this time the termite colony itself is in decline. Gradually, with the death of the queen and, soon afterward, the colony, the mound erodes completely away, and the site returns to its original flatness. Often, a slight concavity is left in which rain accumulates to form a temporary pond. Aquatic plants and seasonal fish somehow reach and then multiply in the pond, but eventually they too yield to the encroachment of the flatland. The site quietly awaits the return of another queen and king termite.

In human history the presence of forts invites the invention of raiders and siege technology. The same is true in the evolution of animals that prey upon the termites. Aardvarks, Africa's premier mammalian termite eaters, use size and the brute force of their powerful forelegs and claws to tear open the walls of the mounds. Their frontal violence, if I may push the metaphor, is the living analogue of cannon fire.

The equivalent of sappers and light cavalry on the other hand are the Matabele ants (*Pachycondyla analis*), named after the fierce Matabele ("Men of the long shield") of western Zimbabwe. The *Pachycondyla* are specialized termite raiders, and built for the task. The black bodies of the workers are exceptionally large for ants, almost the size of a honeybee, and heavily armored. They are also ferocious in battle, even for a bird or mammal that feels inclined to eat one. When I first picked one up at Gorongosa (and afterward vowed it was for the last time), to examine it closely, the ant gnashed its mandibles impressively, then twisted the part of the body behind the waist forward and thrust a long sting into the flesh of my index finger. On a pain index I would rank it just below a hornet. I dropped it unharmed, one of the few ants in my long entomological career to defeat me single-handedly.

●●●

With mandibles full of freshly killed termites, a column of Matabele ants (*Pachycondyla analis*) returns from a successful raid.

A Matabele ant raid on a termite mound is one of the wildlife spectacles of Africa. Now that I have started ranking wildlife, I am obligated to place it well below a lion pride on the hunt, of course, but a good deal above a warthog kneeling on its elbows to grub for roots. A Natabele raid is set off by a single scout that finds and inspects a termite mound within reach of its home nest. I am not sure what the scout is searching for—it may well be a crack in the wall or an exit used by forager termites. If satisfied she (I say "she" because all ant workers are female) runs in a straight line back to her own nest while laying an odor trail. Her arrival brings forth a column of hundreds of nest mates, several ants wide. They run in unison behind the scout, following the trail she has just laid. These raiders are truly the ant equivalent of an *impi*, the Matabele fighting regiment. There is none of the dallying and running around, usually forward but often back in the other direction, that characterizes most kinds of ants. Theirs is an all-or-nothing commitment. Upon breaking into the mound nest, they encounter a phalanx of equally fierce termite soldiers. These they simply run over like heavily armored tanks, crushing and stinging them to death. Then they gather the fallen bodies, along with many of the helpless workers, and march back home.

• • •

A Matabele ant worker carrying a pupa from her own colony.

Local people at Gorongosa are reluctant to drive an automobile or walk over a Matabele ant column. It's bad luck, I was told. Perhaps the feeling is one simply of respect. Certainly I would never want to step on a column in my bare feet—or even with shoes.

On May 11, 2012, while returning during a nighttime excursion in Gorongosa with other biologists, I had the rare opportunity of witnessing an entire Matabele column in another performance as a superorganism. We were being transported by Bob Poole in his elephant-proofed, open-air truck, playing lights to the front and side in search of nocturnal animals. We had just spotted two prowling crocodiles at the edge of the Mussicadzi River when Poole slammed on the brakes, just short of running over a Matabele colony that was crossing the narrow dirt road. This, we quickly saw, was no ordinary raiding party. The ants were at the beginning of an emigration from one nest site to another, pouring in a thick column from the entrance of the old nest, at the base of a small tree, across the road to a temporary bivouac in a clump of grass. There hundreds of workers had already massed, carrying quantities of their immature nest mates—grublike larvae and cocoons containing pupae. From this bivouac a thin trickle of workers was moving back and forth to a hole half an inch wide that led into what must have been a sizable underground cavity. Within minutes the rear guard of the colony emerged from the old nest site. It consisted in large part of small minor workers, a physical caste probably serving inside the nests as day-to-day nurses. There were also many silverfish in the final entourage, tiny teardrop-shaped insects of a species obviously adapted to live with the ants. They acted as though they were members of the colony, running expertly within the column. What was their role? Judging from what I previously knew about symbiotic silverfish, I reckoned them to be uninvited guests that live at little cost to the ants, feeding on bodily secretions and bits of termite prey brought in by the raiders.

Within minutes the bivouacked group began to break up, forming a thick new column that marched to the new nest. The entire emigration was executed with remarkable precision, taking only about an hour. We were lucky to have witnessed most of it.

A Matabele ant colony can defeat a mound-building termite colony in a skirmish but cannot do it serious harm. Its raiders are too few, their number at most a thousand. On the other hand, columns of driver ants, another enemy with immensely larger numbers, can wipe out an entire termite colony during a single raid.

The *siafu*, as driver ants are called in Swahili, exist alongside the mound-building termites as fellow giants of the social insect world. Each colony of at least one of the species contains as many as 20 million workers. The single mother queen is the largest of all known ants, varying in length from 40 to 50 millimeters (about 1½ to 2 inches). Her 14,000 ovarian units are able to deliver 3 to 4 million eggs in a month, exceeding even that of the mound-building termites. The colonies live in soil nests, excavated as an irregular labyrinth of galleries and chambers to a depth of 1 to 2 meters (approximately 3 to 6 feet).

From their underground bunker the driver ant colonies launch raids on an almost daily basis. The ants pour forth by the hundreds of thousands to millions, first as a dense river of bodies, which in the front spread into a broad, forward-moving mass. The front remains connected to the nest by a narrower column of advancing and returning workers, the latter bearing the prey captured by the leaders. The marching raiders are protected by large soldiers, who stand along the edge with their bodies raised almost erect and their sickle-shaped mandibles opened and pointing skyward. Although the ants are blind (why waste energy making eyes if you are going to live shoulder to shoulder in a massed horde?), they have an extreme sensitivity to smell and movement. They are also absolutely fearless, running about and hunting in a frenzy, seizing almost any live animal they encounter. The swarm engulfs all of the ground and low vegetation in its path. In spite of its organized form, it is leaderless. The excited workers rush back and forth at an average speed of four centimeters (less than two inches) per second (slow enough for a human to step away unstung, in case you were wondering). Those in the van press forward for a short distance and then pull back into the mass, allowing new advance runners to take their place. The feeder columns resemble thick black ropes lying on the ground. A close examination reveals them

• • •

Even the hard exoskeleton of a ground beetle is no match for mandibles of hundreds of hungry driver ants (*Dorylus* sp.).

to be dozens or hundreds of workers wide. The ants are so dense that they pile on top of one another and run along one another's backs, while some spill away from the column and form scattered crowds to either side. The frontal swarm, which contains up to several million workers, pinions and kills almost all insects and larger animal life too sluggish to get out of the way. "They will soon kill the largest animal if confined," Thomas S. Savage (he of the appropriate surname) wrote in 1847.

They attack lizards, iguanas, snakes, etc. with complete success. We have lost several animals by them—monkeys, pigs, fowl, etc. The severity of their bite, increased to great intensity by vast numbers, it is impossible to conceive. We may easily believe that it would prove fatal to any animal in confinement. They have been known to destroy the *Python natalensis,* our largest serpent. When gorged with prey it lies powerless for days; then, monster that it is, it easily becomes their victim.

The driver ants enter houses, and the invasion is often welcomed by the inhabitants. The entrance, as Savage reported, "is soon known by the simultaneous and universal movement of rats, mice, lizards, *Blapsidae* [beetles], *Blattidae* [cockroaches], and other vermin that infest our dwellings." In time the ants move on, leaving a vermin-clear and otherwise undisturbed house. One need only wait outside for this service. "When they are fairly in we give up the house," Savage concluded, "and try to await with patience their pleasure, thankful, indeed, if permitted to remain within the narrow limits of our beds or chairs."

The mound-building termites do not await with pleasure the arrival of driver ants at their homes. They are the prey, and as immobile as any well-fed python. The 20 or so kilograms (44 pounds) of the driver ant superorganism are a juggernaut thrown at 10 kilograms of the termite superorganism. Once it finds an opening in the mound, the ants easily overcome the defenders and are able to wipe out all of the colony in a single raid.

Despite the complexity of their social organization and the need of each colony for enormous quantities of food, the driver ants have managed to occupy all of the African continent except for its barren deserts. They have also managed to spread east as far as India and Sumatra. In Africa colonies have been found on the march at elevations where only a few ant species survive. Specialists have been able to distinguish 128 varieties in Africa, a large percentage of which may well be different species. Twenty-one species have been recorded from Mozambique and the countries adjacent to it. At least several occur in Gorongosa National Park, but the exact number remains unknown.

The driver ants have evolved into an adaptive radiation. A few species march through natural fissures and manufactured tunnels underground, while others forage entirely over the surface of the ground. The subterranean raiders operate like military sappers: they can break through into the basement channels and fall upon the termites in a mass assault from the bottom up. Although entomologists have not been able to study the phenomenon directly, they believe that sapper species account for most of the total destruction of mound-building termite colonies. The driver ants that forage aboveground face a greater challenge. On arrival at a mound, they encounter thick, hardened walls that can yield to the claws of an aardvark or a pickaxe wielded by a human, but not to the jaws and tarsi ("feet") of insects. The driver ant columns must find some preexisting opening or crevice in order to enter the nest. Even there they face the miniature equivalent of a Thermopylae: the invasion must be led by a thin phalanx rather than the full front of the driver army, and it may be blocked entirely by the larger force of termite soldiers gathered on the other side. Successful invasions, resulting in the extermination of entire termite colonies, have been observed, but they are evidently rare.

To humans living out their lives in the still unspoiled parts of the African savanna it must seem that nature is at peace. It is in fact the opposite, a dynamic equilibrium built from ferocity. For millions of years the two greatest social forces of the insect world, driver ants and mound-building termites—hunters and prey, warrior Amazons and defenders of fortress cities—have been locked in a slow-moving arms race. Left alone, it will probably continue for millions of years into the future in the balance it created millions of years past. There will be no ter-

• • •

Weaver ants (*Oecophylla longinoda*) do not discriminate against any kind of
prey, even one as well protected as a snail.

mite-versus-ant Armageddon. Neither can completely destroy the other.
It is in human nature to be in an arms race of our own, both among ourselves and with nature as a whole, also built from ferocity. But we have
reached no dynamic equilibrium. We do not as yet even have a sense of
what such a balance might entail. In our case an early Armageddon is
still a strong possibility.

nine

THE LOG OF
AN ENTOMOLOGICAL
EXPEDITION

In May 2012 I led an expedition from Harvard University's Museum of Comparative Zoology to study and classify the unknown ants of Gorongosa National Park. Working out of Chitengo Camp near its south-central border, a short walk to the Púnguè River, our team included Gary Alpert, an expert on the world ant fauna; Piotr Naskrecki, a leading authority on katydids, grasshoppers, and similar insects, with extensive experience in southern Africa; Kathleen Horton, with many years of experience in library research and editing; and myself. Joining us was Jay Vavra, a master naturalist and teacher at High Tech High in San Diego. We were to be accompanied on many trips in the park by the great wildlife photographer Robert Poole. Also participating were Morgan Ryan, here to plan the filming of our *Life on Earth* online biol-

• • •

A portrait of a praying mantis, *Idolomorpha dentifrons*.

ogy textbook (which is being prepared by the E. O. Wilson Biodiversity Foundation), and, not least, Greg Carr.

Thursday, May 10

During the early hours of the night, immediately after arrival at the Chitengo Camp airstrip, we begin looking for ants. Our best hopes are quickly fulfilled concerning the fauna of Gorongosa. Compared with the invertebrate life of a typical north temperate park, the abundance of insects as a whole does not appear notable, but the diversity is extraordinary. Alpert and Naskrecki, while digging holes to set up insect traps, uncover a blind subterranean snake—a miniature reptile resembling a three-inch-long piece of moving thick string. Few people except experienced naturalists would recognize this creature as a vertebrate, but it is far closer genetically to a fifteen-foot python than it is to the earthworm it superficially resembles. Also making their appearance are workers of a subterranean species of driver ant. Unlike most species of their kind, they are red in color, and they move through crevices in the soil along chemical trails.

The high moment of the evening, however, is Piotr Naskrecki's first encounter with Mozambique katydids. He has brought a state-of-the-art ultrasound receiver and recorder, with a capacity of up to 250 kilohertz for recording, and up to 500 kilohertz for detection. Naskrecki has already detected songs of katydids far above the limit of human hearing (which is 20 kilohertz) and even above the ultrahigh clicking of bats during their echolocation. This kind of receiver has opened up a new world of communication and activity in insects. Conventional katydids we can hear unassisted by instruments turn out to be only a small minority of the species in the wild, at least in Mozambique. The full and much larger fauna is now revealed by the detector. The many species are revealed for the first time by wavelength differences in their mating calls.

•••

Entomologist Piotr Naskrecki looking at a sylvan katydid (*Zabalius ophthalmicus*). Also, a conehead katydid (*Pseudorhynchus pungens*) (right), and an armored katydid (*Enyaliopsis petersi*).

The unaided ears of a human walking through the forest at night are assaulted by a riot of unheard katydid cross-talk. Naskrecki is able to go directly to songsters that otherwise would be found only by laborious visual search. Twenty-five species of katydids had been listed as occurring in Mozambique, which Piotr has now increased to about 75. He projects that the actual number is probably between 300 and 350, comparing favorably with the 239 known from the entirety of North America north of Mexico. Among them are not only the common katydid form seen in popular accounts, but also fantastic forms resembling twigs, parts of leaves, and parts of dead tree limbs, all providing visual concealment from predators during daylight hours.

The night brings yet another kind of adventure. Alpert, Naskrecki, and Vavra call us to witness the aforementioned swarm of driver ants in progress inside Chitengo Camp. The workers in a colony of this species number in the hundreds of thousands, and during a swarm raid a large fraction of these pour forth from their underground bivouac as a swiftly moving column, which upon extending begins to divide in smaller branches to create a widening fan, then still smaller rivulets, and finally a field of ants hunting singly or in small groups. The horde covers every square inch of an area ten meters (thirty-three feet) in breadth. Walking carelessly into the outer fan, we soon feel ants crawling up and stinging our legs. Their attacks are annoying but not injurious. We find none of the soldiers with fearsome sickle-shaped mandibles that some driver ant species have. Then at last, bedtime after a hard day of travel and facing a big day tomorrow.

Friday, May 11

With Bob Poole driving, we travel several kilometers into the savanna and dry woodland and stop at a fever tree grove growing on a swale of slightly lower, moister soil. This habitat, as we hoped, proves to be especially rich in ants. Gary, in particular, is able to collect workers of four species foraging separately on the trunk of a single tree. (The name fever tree, incidentally, came from the realization by explorers and early settlers that malaria is more prevalent wherever this particular tree grows—due, as we now know, to the greater abundance there of malaria-carrying mosquitoes.) We all manage to harvest many more species, belonging to several genera. The most notable discovery is a colony of the giant ponerine *Platythyrea cribrinodis*. Carefully avoiding their famously painful stings, we pluck up workers for our collection. We also get to examine a huge *Nephila* orb-weaving spider suspended in a six-foot-long web. Brilliantly colored, she sits immobile near the top. Huddled beneath her body like a swaddled baby is her ludicrously small mate. Elsewhere on the web we find an even smaller, silver, globular theridiid spider, a species that lives on the long strands of the *Nephila* web and between which it constructs smaller webs of is own.

From his first day's search here and elsewhere, the fast-moving Piotr Naskrecki finds twelve species of praying mantises, a number he says is "off the chart." No one disagrees.

That evening we drive to the edge of the Lake Urema floodplain grassland to view herds of waterbuck, impala, wildebeest, and warthogs. As the sun sets, flocks of spur-winged geese, two kinds of ibis, and other wetland birds pass overhead on their way to their evening roosting sites. Undistracted by the spectacle, Naskrecki listens to the ultrasound songs of yet other species of katydids in the grassy sward leading down to the lake. I persuade everyone to follow the final descent of the sun past the western horizon, hoping for a glimpse of a green flash. We don't see one—no surprise, since I have failed hundreds of times viewing other sunsets around the world.

As we bump along on our way home through the dark forest, Bob Poole suddenly slams on the brakes of his elephant-resistant vehicle, just missing a column of Matabele ants crossing the road. We all jump

out to watch this event. This time the large black ants are not raiding a mound-building termite nest, but instead are emigrating from one nest to another (as I described in an earlier chapter). Over the next hour we are able to witness almost the entire process. Naskrecki records the audible clicking sounds made by the ants as they run.

● ● ●
Yellow-billed storks of the Gorongosa floodplains.

● ● ●
Some of the praying mantises found in Gorongosa (clockwise from the top): grass mantis (*Epitenodera capitata*), boxer mantis (*Otomantis scutigera*), cryptic mantis (*Sibylla pretiosa*), and flower mantis (*Pseudocreobotra wahlbergi*).

The next morning Piotr shows us the remarkable result. The Matabele ants make two sounds: a low-frequency click that we can hear, and may have the purpose of warning off predatory birds and mammals, topped in each pulse by an ultrasound segment quite likely a means of keeping the fast-moving column in a tight formation—fife and drum, so to speak. This new phenomenon, I point out, deserves further study and publication in a scientific journal. Most of the group then goes to a distant grassland near where lions have been seen. The effort yields no lions but instead four genera of ants not collected previously. I stay behind at Chitengo Camp, sorting the material already collected and preserved in alcohol, then examining each insect under a microscope to identify the species. I'm able to take specimens at least to the level of the genus to which each species belongs. The tally to date is 21 genera, containing 55 species. Thus in just three days of effort in the exuberant Mozambique fauna, we have a list of genera and species longer than that of the entire British Isles, and at this rate we are closing fast on New England.

Sunday, May 13

More sorting and identification of specimens. Gary and Piotr depart for more collecting.

Monday, May 14

The group travels by an open vehicle to the location where a pride of lions had just been seen. We are accompanied by Bob Poole and Paola Bouley, the latter a young woman conducting research on the status of the lion population at Gorongosa National Park. The key problem needing research at this time is why the population is increasing more slowly

than expected, despite the abundant food supply and apparently excellent physical condition of the lions. We find our quarry in the shade of a copse next to the road. It is a pride composed of a young male and three young females. They lounge in an early afternoon siesta, their bellies full, apparently unconcerned with the six humans staring at them thirty feet away. There is something about lions that cancels distraction in all other directions. One of the females walks casually out of the copse, comes partway to the vehicle, defecates, and returns. A gesture of contempt? I

• • •

A Gorongosa lioness.

don't think so. In any case, Paola is excited, anxious to pick up the scat for DNA sequencing, because then she can associate the genome with a particular recognizable individual. Of what use is this information? She needs data to test the hypothesis that the small Gorongosa population is forced to inbreed and hence suffers loss of fertility. At first she is frustrated: the scat is too close to the lions for us to risk getting out of the car to collect it. Poole maneuvers the car carefully between scat and lion, and Paola picks up the scat in a plastic bag. As she prepares to put it away, a small insect flies off it. "What is that bug?" she asks me. An entomologist, I'm able to explain that it is a scarab beetle that has come to collect a bit of the dung on which to raise its young. Good fresh scat is hard to get. The beetle beat us to the prize, but no problem; its presence won't spoil the DNA sequencing.

In the late afternoon we set out with Greg Carr for the distant "Explore Gorongosa" camp, where visitors can stay in tents in the midst of woodland next to a stream. On the way, we witness an event extraordinary even for Gorongosa. At the left side of the road in high grass at the foot of a tree, an eagle is struggling to lift a black mamba and fly up and away with it. The snake is one of the deadliest venomous species known in Africa. This one is a big adult, seven feet long as best we can estimate (no one volunteers to get out and make sure), with a thick, powerfully muscled body. Disturbed by the car's approach, the eagle drops the snake and flies away. The mamba then does the opposite of what might be expected, which would be to slither away and hide in the thick vegetation. Instead it climbs the tree, laboriously and conspicuously, to reach the highest branches. Then it proceeds to wind its way around in the canopy, and is still there when we drive away fifteen minutes later. Whether the eagle returned to try again, this time enjoying its prey in a more vulnerable position, we'll never know.

●●●

The limestone gorges of
the Cheringoma Plateau
of Gorongosa are a
little-explored treasure
trove of biodiversity.

We take a journey to a lost world, comprising a unique ecosystem with an unknown fauna. By helicopter we fly low northeast from Chitengo Camp, past Lake Urema, leaving the Gorongosa massif a dark rim on the western horizon, on across a savanna wilderness, its low trees separated by seasonal grassland and bare white pans. It seems that this classic African landscape, from the Great Rift Valley onto the Cheringoma Plateau, might go on forever. But suddenly there is a series of dips in the terrain, each first visible at a distance as slashes of green, then appearing as a stream with lush vegetation lining its banks. At the end we come to a deep gorge that shelters the little Ngagutura River, made invisible by a closed forest canopy. The dominant tree along the rim, we learn, is Lebombo ironwood (*Androstachys johnsonii*), a shrub widely distributed in southern Africa. The sides of the gorge plunge steeply down, in some places as vertical granite walls. At the western terminus is a bare limestone space at the summit of the scarp on which the helicopter can land. At the edge of the space, the view is straight down for more than a hundred feet to the top of a fully developed tropical African rain forest.

This oasis of lush tropical growth in the midst of the arid African savanna has been visited rarely—except by local native hunters. It can be reached and descended only with great difficulty on foot, but fortunately the trip lasts less than an hour by helicopter from Chitengo Camp. In October 2007 researchers from South Africa had come here as part of a survey of all the trees and shrub species of Mozambique. In their first preliminary and incomplete sweep, they found eighteen plant species in the depths of the gorge, of which only one, a fig (*Ficus ingens*), also occurs among the Lebombo ironwood, plus forty other species on the lip of the gorge. At the bottom they found a second fig, *Antiaris toxicaria*, the first record of this species south of the Zambezi River. The botanists noted an abundance of orchids growing on the trees, one of which is a virtual air plant in form, absorbing all its moisture through free-hanging, spidery roots. The plant life overall is exuberant. Even the limestone faces are home to maidenhair ferns, a species of *Adiantum,* accompanied by the dry-adapted *Ficus ingens.*

Our descent occurs four and a half years after this preliminary botanical survey. We are the first entomologists into the gorge. We are excited by the knowledge that with the South African botanists, we are now the biodiversity pioneers of such an unusual environment. Fortunately, we can make it all the way to the bottom in the helicopter without rope climbing. But it is a complicated process. First, the pilot, Mike Pingo, lands on the bare outlook, then ferries us and our equipment in two trips to the bottom, settling down on a grassy clearing made the previous day by park staff, also transported by helicopter to the area. Six of us enter the forest together: Gary Alpert, Piotr Naskrecki, Jay Vavra, Bob Poole, Greg Carr, and myself.

Although there are snares and other signs of hunting in the surrounding savanna, there is no evidence that the forest of the gorge has ever been disturbed, except by paths cut by poachers to obtain water from the bottom stream. So far as we can see, the limestone gorge is still almost completely free of human disturbance.

Here then is the general scientific problem addressed by assessing the fauna and flora of this remote and unexplored habitat. The area of the rain forest is relatively small, making the likelihood of any particular species going extinct each year relatively greater than is the case for similar forests of larger extent. The distance to other rain forests from which additional species might immigrate is also substantial. Both of these conditions, small size, great distance, add up to an expected lower number of species present in the rain forest of the limestone gorge. On the other hand, the rain forest is probably very old. According to geologists, caves around the gorge were filled partly by marine deposits 3 to 20 million years ago, then eroded out again 0.5 to 1.5 million years ago. Even if the caves and gorge have persisted only since the newer erosion episode, there has been time enough to permit the origin of plant and animal species unique to them.

So, as evolutionary biologists we asked the following question in advance of our visit: would we find a relatively small number of ant species inhabiting Gorongosa's limestone gorges, but some of which are found only there? Even allowing for the brief period on a single trip we were able to work there, and at a single spot in the gorge, we discovered a surprisingly large number of ant species. Some also occur in the surrounding savanna and dry forest, but a few appear to be unique, at least still unknown elsewhere. A definitive answer to the question raised by biogeographic theory cannot be given right away. It will come only after a great deal of additional research is conducted in the field and museum, with the aid of a more comprehensive map of ant distributions in southeastern Africa.

Wednesday and Thursday, May 16–17

This two-day period we spend sorting and labeling our collections to date. I stay at the microscope, identifying species as the stream of ant specimens come in. Gary Alpert and Jay Vavra man the Winkler traps, which separate still more specimens from the soil and leaf litter brought from the collection sites. Kathleen Horton looks to the records and storage of specimens. Our specimens now number in the thousands (no surprise; they are ants), and I'm beginning to get wobbly on some of my identifications (they will be checked later). But we are keeping up the pace from field to labeled vials. Some of the species will undoubtedly prove new to science. To be sure, closer studies will need to be conducted in the ant collection back at Harvard University. One exciting find has been a specimen of the primitive termite-hunting ant *Promyopias silvestrii,* the only species in its genus known anywhere and extremely rare. Only twenty other specimens have ever been collected, and ours is the first record from southeastern Africa.

Friday, May 18

Today we travel by helicopter to the summit rain forest of Mount Gorongosa. This habitat is completely isolated from other mountain rain forests and has had a long period of geological time to evolve a distinct ant fauna. It is the altitudinal reverse of the Ngagutura gorge forest. We have no idea what to expect. The previous August I had visited another part of the forest briefly at about six thousand feet elevation, and found—to my disappointment—only four species of ants. The commonest one was a colony of driver ants, which may, I surmised, have destroyed the other ant species. That would account for the small amount of ant diversity. Today I hope to discover a much larger array of species, including perhaps some new to science. We land in a beautiful meadow at 5,300 feet elevation, close to the edge of forest and well away from my 2011 site. Our hopes fade quickly. Hard work in the meadow and deep into the forest yields only eight species, each relatively scarce. They include driver ants. What is going on here? Are the montane woods cursed by invasions of driver ants that drive everything else to extinction? I don't think so, not anymore. At the end of the day we conclude that the real cause is the relatively cool, wet environment, conditions that are poor for ants generally. We are apparently at the upper elevational limit of ants in this part of the world. And the driver ants? They survive because they are able easily to migrate up and down, into the high forests and out again, whereas other ants are rooted to the place where their colonies are founded, and they mostly perish.

Saturday, May 19

The group splits up, Gary to collect ants, and more ants and still more, Piotr to collect orthopterans with equal industry, Kathleen to catch up on our still-heavy correspondence. I join marathon conferences, first with a newly arrived BBC crew to help plan the final episode of a five-part series on African conservation, called *Africa,* with Gorongosa featured (they tell me), and later on the same day with the *Life on Earth* team to discuss the last of three chapters to be filmed at Gorongosa for our online introductory biology text.

Sunday, May 20

Not a day for rest. There never is during fieldwork, with so many exciting discoveries filling every day. Our collecting of ants and orthopterans continues. Most notably, Gary Alpert and Piotr Naskrecki encounter a grove of fever trees through which elephants had passed an hour or so previously, knocking down some of the medium-sized trees. The opportunity presented itself to collect ants living in the canopy, so Gary and Piotr, keeping their eyes open for elephants, work through the newly accessible upper twigs and branches. They discover several arboreal species not previously seen.

● ● ●

Fever trees knocked down by feeding elephants allow entomologist Gary Alpert to collect ant species found only high in the canopy. More than two hundred species of ants have been recorded from Gorongosa, including (clockwise from the upper left): driver ant (Dorylus sp.), Pachycondyla tarsata, big-headed ant (Pheidole sp.), tumbling ant (Melissotarsus emeryi), Myrmicaria sp., trap-jawed ant (Anochetus levaillanti).

●●●

The "lion's head" is a natural growth on the trunk of a giant baobab. *Photo by Kathleen Horton.*

Monday, May 21

The helicopter takes me, Kathleen Horton, and Tonga Torcida to a woodland dominated by mopane, a leguminous tree with exceptionally hard wood and a dwarf canopy only about ten to twenty feet in height. The soil and thin litter are bone dry and sprinkled only with sparse ground plants—yet teeming with ants. We rendezvous with the BBC film crew to add to the final episode of the *Africa* special. They arrive in vehicles carrying heavy photographic equipment and film Tonga and me as we collect ants, so double duty is served—acquiring a lot of specimens and having a film made of the collecting process. Afterward we drive less than a mile to visit an immense and ancient baobab, quite likely the largest in Gorongosa National Park, and due to its remote location previously seen by few visitors. In the center of the massive trunk is a gnarled outgrowth shaped with the astonishingly close likeness of a six-foot-long lion's head facing forward, almost perfect: left ear, face, nostril, mouth, and chin. It could have been sculpted by human hands, but close inspection confirms that it is indeed natural. Arriving back at Chitengo Camp, I tell Greg Carr that if a road is ever built into this part of trackless savanna, the tree would be a prime tourist attraction and camping site. I'd call it Lion's Head Baobab Camp.

Tuesday, May 22

My morning is spent filming an interview for *Africa*, making connections between the unique qualities of Gorongosa National Park and the general importance of national parks worldwide. Hard work. But it was more than compensated by knowing the potential size of the television audience and the rich wildlife diversity at the interview site, on a bank of the Mussicadzi River close to Lake Urema. Ground hornbills, two species of ibis, a fish eagle, crocodiles, warthogs, and waterbuck were all around. On the way back to Chitengo Camp we stop for a glimpse of a fifteen-foot (or so) crocodile seen recently. We may have glimpsed it in the water a hundred yards away. Gary has begun an intense collection of ants in dry forest on the road from Chitengo Camp to the Púnguè River. Among the species he's found is a tiny ant so distinctive that it may be not just a new species but a new genus. (Later we learn it is a species of *Cerapachys*.) Jay and Morgan make a helicopter tour of the entire park to photograph some of its fifty-four classes of ecological habitats from the air.

Wednesday, May 23

Most of us scatter into different sections of the park, variously by foot and automobile, to collect yet more specimens. My own effort on discovering and classifying the ant fauna (we are now up to thirty-five genera and more than one hundred species) is slackening off rapidly as I prepare notes and silently rehearse for lectures to be filmed for the *Life on Earth* course, which will be available via iPad by our partner, Apple.

Thursday, May 24–Tuesday, May 29

During this last week of my visit to Gorongosa, I find myself immersed in preparation of three of the chapters of *Life on Earth,* devoted, respectively, to human evolution, animal behavior, and conservation biology. Each requires my composing lectures on these subjects, then delivering them at a meaningful locality in the Gorongosa forest. Meanwhile Gary Alpert and Piotr Naskrecki continue their explorations of the ant and orthopteran faunas.

Wednesday, May 30

Kathleen Horton, Jay Vavra, and I depart for our various destinations in Massachusetts and California. Gary and Piotr remain at Gorongosa for another month to continue collecting and add to their natural history studies.

•••

It is an oft-stated rule among taxonomists and biogeographers that in order to complete a collecting expedition properly, each day in the field should be followed by ten days at home in the museum or laboratory. The ant collection housed in Harvard University's Museum of Comparative Zoology, to which the Mozambique material was delivered, is the largest in the world, comprising an estimated seven to eight thousand species represented by, perhaps, a million specimens ready for research. Preparation of new material is not a simple matter, and it is labor-intensive. Each specimen is removed from the alcohol-filled vial in which it was placed in the field. It is dried and glued to the tip of a paper triangle, which is pierced in turn by a thin pin. A label, small in size and in lettering, is also pierced by the pin and denotes the locality, collection, date, and name of collector. A bar code may be added that will reflect the information on the labels, including habitat, nest site, and other details. A second label is added that contains the scientific name of the specimen—if that is known. This label has a special color and notation such as "holotype" (the reference specimen on which the description and name are based) or "paratype" (another specimen of the series placed in the same species as the reference specimen). Types are extremely important in taxonomy; they serve as the ultimate reference in case of dispute over which scientific name should be applied to all specimens of the same as opposed to similar species.

Museum specimens are prepared this way to provide easy access and examination under microscopes. You just pick the pin up by the beaded head and stick it in a rotating mount on the microscope stage—where it can be examined from different angles and at different magnification with a twist of the fingers. The specimens are stored in trays with others of the same species. Incidentally, the oldest ant specimen I've examined, borrowed from the Museum of Natural History in Paris, was a worker ant collected in Colombia in 1826.

Our 2012 ant samples from Gorongosa have entered the collection at Harvard. We hope they will be there in perpetuity for the use of future generations. In time, the specimens will be divided, with some deposited in a reference collection planned for Gorongosa National Park and others in the Mozambican Natural History Museum, at Maputo.

ten

THE STRUGGLE FOR EXISTENCE

On a late summer afternoon in Chitengo Camp I came upon a velvetberry bush in full bloom. Its spikes rose to head level and were crowded with blue florets. Around these miniature blossoms fluttered a hundred or more pierid and nymphalid butterflies of medium size. There were about a dozen species represented, all predominantly white in color but distinguishable by their wing margins, which were variously striped, checkered, or tipped in orange and black.

I had no idea where these butterflies came from or where they went upon leaving. I was just immersed in the moment, beguiled by the beauty conjured before me. I stayed there for a while, and came back the next day, and the next. After a while and for no particular reason I asked, What are these insects thinking? Is it anything more than the flower and nectar? Do they feel in their six-legged way any of the beauty and drama of the spectacle they create?

From insect physiology and neuroscience, of course, I knew the answer: the beauty and drama and other emotions that brought me to Mozambique and this velvetberry bush were entirely in my own head. In the

● ● ●

In a rare role reversal an invertebrate predator catches and kills a vertebrate prey.

tiny brains of the butterflies there was only recognition, arousal, and action. None paid any attention to me until I brought my fingers close enough to grasp their wings and pull them off the florets for closer examination. What to do about an approaching human is not encoded in tiny butterfly brains.

Biologists have come to understand very well what insects like these actually see, smell, taste, and touch. All of it is radically different from our own senses. Their eyes are composed of scores of hundreds of ommatidia, each a receptor of a whole image from a slightly different angle than all the others. Such a compound eye, as entomologists call it, may seem a very unwieldy way to form a picture, but it is just what an insect needs: supersensitivity to sidewise movement and from that the ability to detect the approach of a predator or prey. If you wish a demonstration of this principle, try walking up to a dragonfly perched on a grass stem or snatching a housefly off a tabletop with your hand.

While I'm at it and because it has some relevance, I'll tell you how to "hypnotize" a dragonfly. The method was taught me by a fisherman I recently met on the Tensaw River in Alabama. Come as close as you can to a dragonfly sitting on a perch, without causing it to fly off. Moving carefully, raise a hand to your chest and start wriggling your fingers at a moderate rate, not too fast as typing a letter, and not too slow, as molding clay. Then gradually (this may take some trial-and-error practice) move the hand away from your body and out toward the dragonfly. To your surprise, if successful, you will be able to reach within several inches of the sitting insect. Now (carefully!) push your index finger to beneath its legs and slowly lift. The dragonfly will then climb aboard your fingertip and use it as a new perch. Have you really hypnotized a dragonfly? No, you've just tricked it by imitating leaves shifting about in a breeze. You've proven, if it matters, that you are smarter than a dragonfly.

Through science we can speak of what insects like dragonflies know by instinct. But we may never be able to travel through the neuronal circuits of their brains in order to perceive from the inside the nature of whatever consciousness they possess.

For insects and other small animals life is hard, by which I mean overwhelmingly, minute-by-minute hard. Survival is possible only by precise navigation on the tricky, narrow roads of instinct. The velvetberry bush at Chitengo Camp was not a thing of beauty for the butterflies; it was an arena of life-and-death competition. Not enough florets were there for all the butterflies to land at the same time and feed at will. Many of the little blue flowers, evidently drained of their nectar, were attended only briefly. The larger species of butterflies repulsed the smaller ones whenever the two came closely together, and the larger were able to stay on each floret longer after landing. Only their relative scarcity in the whole population of butterflies kept them from completely dominating the velvetberry bush.

Darwinian success in such aerial contests is the result of struggle and luck. Winners on the bush filled their stomachs more quickly and departed the flowers after feeding to locate and deposit eggs on the kinds of food plants to which their species are adapted. Losers were more likely to stay longer at the bush and perhaps were forced to leave hungry—with less ability to lay eggs.

All the butterflies, whether well fed or not, were in peril from enemies stationed around the velvetberry bush. On nearby shrubs, orb-weaving spiders hung motionless in their webs, waiting to snare and bind in silk butterflies that chose the wrong flight path traveling to and from the bush. Jumping spiders, eight-eyed and eight-legged tigers of the arthropod world, prowled the ground and climbed the lower foliage, ready to pounce on butterflies that came within reach. Crab spiders and praying mantises, with camouflaged bodies mimicking variously flowers, twigs, or leaves, sat motionless in the vegetation, intending to snatch victims from ambush. The spiders used long, spread forelegs to pull prey to their waiting fangs, and the mantises worked with spiny forelegs that swung open and closed like switchblade knives. On the ground, wolf spiders and carabid beetles patrolled for butterflies that dared to alight nearby, while an occasional robber fly descended from above to seize those that kept flying. Every move the feeding butterflies made, every decision to stay or leave, every blossom approached or avoided had huge consequences for these insects. Every choice was a matter of life and death.

I didn't see most of the butterfly killers right away. I wasn't supposed to. If they stood out in shape or color to me, they would have also been the more easily evaded by their intended victims. The predators would also have increased the chance of falling prey to one another, as well as to the ever-hungry lizards and birds that formed the next level up in the carnivore food chain.

It is in the nature of things that not just prey but also predators live on the edge. The butterflies and other prey they hunt are among the restraining forces that limit the growth of the predators' populations. As such they are the equivalent of floret nectar for the butterflies. Both assemblages of insects illustrate the same law of ecology: if a species anywhere in the food chain acquires a bonanza of food, enough to fill all stomachs, individuals of that species will reproduce more and as a result the population will increase. But so will the populations of their predators, which have thereby been provided a bonanza of their own. Predator and prey populations, such as those respectively of crab spiders and butterflies, pull against one another and bring their populations back to the status quo ante. Each creates tighter feedback loops within the larger ecosystem of butterflies, spiders, and all the other interconnected prey and predators around them. The whole flower-butterfly-predator food chain is part of an even larger ecosystem of these and other kinds of feed-

●●●

In the struggle for existence
even the strongest wings
may not be enough if you
stray too close to a well-
camouflaged predator,
such as this cryptic mantis
(*Sibylla pretiosa*).

•••

Lappet moths (*Chrysopsyche lutulenta*) emerge at the beginning of the dry season, when leaves of deciduous trees turn yellow. Their seemingly conspicuous coloration allows them to disappear among the fallen foliage.

back loops. The bumpy equilibrium of numbers all this produces over time is what we call the balance of nature.

It has become rather louche in educated discourse to speak of nature as red in tooth and claw. However, let us not evade the plain facts in this matter. Nature *is* red in tooth and claw, and that is the case all the time. Still, casual visitors to wild environments, like those at Gorongosa, cannot expect to witness evidence of this verity within a period of mere hours or even days. As you walk at leisure along a lakeshore or through a forest you may see a heron spearing fish, or swallows snatching insects on the wing, but you will otherwise rarely witness a birth or death, much less the rise and decline of entire populations. To sense the struggle for existence and the deadly harmony of an ecosystem within which it unfolds in a relatively short period of time, it is better to stop for a while and gaze for a few hours at a flowering bush or several square yards of bare ground. The organisms that become apparent are not the large, conspicuous mammals and reptiles that crash away from you through the undergrowth or noisy birds flying out of sight through the canopy. The animals to watch are the little things abounding on twigs and leaves and running about in the leaf litter and humus. They live more harried lives in a very different scale of space and time than you and I.

Homo sapiens, one of the biggest of all animals, is a species native to the African fauna. We originated there, and it is natural that we should be attracted to its dry forests and savannas and animals, the ones to which our ancestors were intimately connected throughout most of the last million years of evolution. But the condition of walking on two legs lifts our heads three to six feet above the ground, an immense distance from the tiny insects and other invertebrates that make up the bulk of animals in these ecosystems. So while we belong in such an environment by virtue of our deep genetic heritage, we are not built to be good naturalists. Our perception of this or any other wildland is intensely anthropocentric. Our hunter-gatherer genes predispose us to focus on what we can eat, hence vertebrate animals, the larger the better, along with nuts, tubers, succulent vegetation, and fruit whose conspicuous colors among green foliage advertise their ripeness. Most compelling of all the organisms to which our senses come jangling alive are the few large animals capable of eating us.

We are so drawn to big herbivores grazing on the plain and the dangerous carnivores stalking them (as we ourselves once did) that we make long, expensive trips just for the thrill of coming to this land, our ultimate home. Our love of nature is in our genes. Instinctive and bubbling with poetry, that love is nevertheless intensely selective, sifted and programmed by the ancient necessities of daily survival. This part of human nature spills over to favor domestic animals we have gathered about us, but only insofar as they serve us as assistant hunters, guards, status symbols, and child surrogates.

We may think these favored species share our feelings. Some come close—dogs, the ultimate pack animals, and chimpanzees, our closest animal cousins. But the vast majority of animal species live in a sensory world alien to our own. Their brains have been put together differently. They were sculpted by evolutionary forces of another kind than our own.

To understand nature at any depth beyond a walk in the woods is to think upon the sensory world of animals. Only a relatively few species, including birds, clearwater-dwelling fishes, and nonhuman primates, are primarily audiovisual as we are. Almost all the rest live by some combination of touch, taste, smell, electrical pulses, polar magnetic fields, ultrasound, and polarized light. And that is just the beginning of the strangeness of the natural world. In order to understand any animal species thoroughly, you must also know its life cycle, food requirements, symbionts, enemies, and, ultimately, something of the evolution that locked it into the niche in which it survives today. Biologists themselves have accomplished that much in fewer than one species in ten thousand. The disciplines of ecology and evolutionary biology still have a dismayingly great distance to go. Given that we have scarcely begun to understand the origin and meaning of our own species, how can we hope by any easy means to master the rest of the living world?

But that aside, let me at least make clear that the members of each species are the best of the best, including those poor butterflies at the velvetberry bush dodging enemies on all sides. They are survivors and descendants of survivors that have won their struggle for survival in test after test across countless generations. In some birds and mammals, one in every two to five or ten individuals born lives to reproduce. In leaf-cutter ants, one in ten thousand fertile queens that depart from the mother colony succeeds in starting a colony of her own. And in the ocean, representing the most extreme cases in all the world, only one larva among millions in some pelagic species makes it to maturity.

The balance of nature in every ecosystem is thus an equilibrium teetering on a razor's edge. Even small changes in the environment can tip it enough to extinguish species. The golden toad of Costa Rica was an exquisite bright yellow animal that had lived for ages in continuously moist mountain rain forests of Costa Rica. In recent times its species was confined to a ten-square-kilometer (3.8-square-mile) patch of forest in the Monteverde reserve. Recently the mountaintops dried somewhat due to climate change, introducing for the first time a brief dry season. The golden toad may also have been struck by the deadly chytrid fungus. The species fell in numbers swiftly, and by 1990 it vanished forever. In parallel manner the famous ivorybill woodpecker of the southern United States specialized on large, old cypress and other trees to build its nests

and provide the beetle grubs on which it fed. When the big trees were cut for timber, especially in the floodplain forests, the bird plunged to extinction. Although sporadic sightings of ivorybills are still reported, the last one confirmed was in 1944.

The loss of a single tree species often results in the elimination of multiple moth, beetle, and other insect species dependent on it. Similarly, the disappearance of a key pollinator can threaten multiple plant species. When honeybees recently suffered a die-off, probably from a combination of pesticides and inbreeding, some crops dependent on their services declined with them. When such symbioses are tight and highly specific, as between some tropical orchids and euglossine bees, the extinction of one partner means doom for the other.

Disease organisms are especially dangerous. The establishment of a single alien pathogen of uncertain origin can be catastrophic to the new hosts it invades. The accidental introduction of a chytrid fungus (possibly African) into Australia, New Zealand, Europe, Africa, North America, and the American tropics has resulted in the loss or reduction of scores of species of frogs. During the past one hundred years, human modification of rivers, streams, and lakes in North America alone (mostly through damming and pollution) has caused the extinction of at least sixty freshwater fish species. Our environmental domination of the land and sea has reduced Earth's biodiversity through the extinction of at least 10 percent of its plant and animal species, mostly in the last century. It is driving species at an accelerating rate through conservation biology's descending categories of "threatened," "endangered," "critically

• • •

Caterpillars of the owlet moth (*Rhanidophora ridens*) advertise their unpalatability with gaudy coloration and plumelike hairs.

endangered," and, finally, "extinct." If this perception seems overstated, consider one of the soundest principles of ecology: the number of plant and animal species sustainable in a habitat increases with the area of that habitat (or decreases with loss of it) by roughly the fourth root of the area. When, for example, the habitat is reduced in area by 90 percent, the sustainable number of its species is cut in half.

Some of the most biodiverse habitats in the world have been reduced by 80 to 90 percent of their areas, including the rain forests of Madagascar, the Philippines, and the Atlantic coast of Brazil. It would require only a single careless (and uncaring) step to remove the remaining 10 to 20 percent. That conceivably could happen if the land were considered of vital importance to agriculture, mining, or the spread of growing human populations. In such an extreme case, the sustainable number of species limited to the rain forests would plunge from half of the original to nearly zero.

Humans come first, of course. But shouldn't the rest of life and the quality of human life dependent on the rest of life be entered into the equation? Put another way, do we wish future generations to think we were insane or perhaps criminally stupid?

The grim reapers of biodiversity, all born of human activity, can be clustered into five categories, conveniently combined in the acronym HIPPO (like the animal). The letters, in order of appearance, and from most to least importance, are *H* to represent habitat destruction, including that caused by climate change; next, *I* for invasive species, such as the chytrid fungus killer of frogs; the first *P* for pollution, a principal factor, for example, in the extinction of one-third of the fish species in China's Yellow River; the second *P* for continuing population growth and the acceleration of consumption that accompanies it; and, finally, *O* for over-

harvesting—the removal of the last individuals of a species by relentless hunting or fishing.

The impact of invasive species, defined as alien species that cause some amount of harm to the environment or people, is especially insidious, because it is often detected in the early stages only by experts. Hawaii, the most egregious American example, appears to newly arriving visitors to be a lush, biodiverse paradise. But virtually all of the plants and animals they see are aliens, introduced from other countries. The native Hawaiian birds, down to two dozen species from an original 120 or more prior to the arrival of humans almost two thousand years ago,

•••

Native to Gorongosa but accidentally introduced to other tropical and subtropical areas of the globe, the big-headed ant (*Pheidole megacephala*) has become a serious, invasive pest. Here they have overpowered and killed another Gorongosa native, a male driver ant (*Dorylus* sp.).

hang on only in fragments of remote, cold mountain forests. The original Hawaiian plants are in similar peril, with extinction already extensive and a few of the surviving species known only from single specimens.

It might be thought that Hawaii nevertheless acquired a beautiful immigrant flora and fauna, even if they replaced the original native species. But the obliteration of the magnificent original fauna and flora is only part of a catastrophic loss occurring everywhere. Most of the plant and animal species introduced into Hawaii occur as aliens in other sites around the world. In a few cases local biodiversity may be increasing by the buildup of these invasives—but global biodiversity is decreasing to a corresponding degree.

And so it has ever been, wherever and by whatever means humanity has altered the environment for its own benefit. The effects are often subtle, at least at first. For example, as the remaining rain forest wildernesses, located in the Amazon-Orinoco and Congo basins and New Guinea, continue to be cut back and crisscrossed by new roads, the edges of the remaining forest actually gain in numbers of species. These new habitats, called ecotones, support fragments of the original vegetation plus grassland and specialized scrub vegetation. But nothing has been gained in overall biodiversity. Just as in the gardens of Hawaii, the same species dominate the ecotones over wide areas throughout the tropics. The loss is deepened by a second kind of edge effect: due to increased sunlight and wind currents, the uncut forests grow drier for as much as one hundred meters (328 feet) inland, reducing the variety of species limited to deep-forest conditions and, with the increase of intrusion, ultimately extinguishing them altogether.

We must accept that while its practical goals have become crystal clear, the science of ecology is still mostly a shadowed book of mysteries. The struggle for survival of butterflies on a flowering bush, creating the delicate balance of their numbers, magnified to the ecosystems to which they belong, is a part of nature still too complex for science to fully comprehend. Until that level of understanding is more closely reached, it will be wise for humanity, in the spirit of the precautionary principle, to move as slowly and carefully as possible, if at all, into Earth's surviving wildlands.

●●●

A new season requires a a
new home and, at the onset of
the rains, male village weavers
are hard at work building
intricate nests.

eleven

THE CONSERVATION OF
ETERNITY

Communication and the world economy have been largely digitized. The planet is enmeshed by the Internet, and more and more of the world's great store of knowledge is instantly available to anyone, anytime, anywhere. Populations are imploding into cities as most of humanity continues to move off the land. By 2012, a dozen cities exceeded ten million in population. One of the largest, Karachi, Pakistan, would if it were a country exceed Greece or Hungary or Portugal. In the faraway Chitengo Camp of Gorongosa National Park, staff members from nearby villages chatted over cell phones. The once-forbidden summit of the sacred mountain was only thirty minutes away by helicopter.

Since I was born, eighty-four years ago, the human population has grown threefold. The wildlands of Earth are shrinking accordingly, in area and in the human mind. Nature remains something out there, remote, a caricature on a television screen, expendable, going, going, gone—folded into the cost of an oil field or timber tract.

Many writers call this new age the Anthropocene, wherein the planet is of, by, and for humanity. In the worldview of the corporate priesthood, the restructuring of Earth to accommodate vast numbers of people and their artifacts is not the price of progress. It *is* progress.

To those who feel content to let the Anthropocene evolve into whatever destiny we are about to stumble on mindlessly, allow me to make a modest suggestion. Don't stop, but please set aside the largest fraction of Earth's surface possible as inviolate nature reserves. One-half would be nice. Conserving this land and sea is noble and a workable goal, especially when it includes those sections with the greatest density of species. I believe that the ten billion people expected to be present at the end of the century will enjoy a far better quality of life if we conserve half of the planet for nature than if we consume nature entirely.

How would natural parks and other reserves be preferable to a dozen more megacities? Let me count the ways. Wildlands maintain the stability of the physical world. They are our birthplace. Our civilizations were built from them. Our food, most of our dwellings and vehicles were originally derived from them. Our gods once lived in them. Nature is the birthright of everyone on Earth. The millions of species we have allowed to survive are our phylogenetic kin. Their long-term history is our long-term history. Despite all our fantasies and pretensions, we always have been and will remain a biological species tied to this particular biological world. Millions of years of evolution are indelibly encoded in our genome. The genes plot the cellular circuits of our brains; they play out in the tapestry of our minds and fingers. To let more of Earth's biodiversity—perhaps we should say more simply the rest of life—continue its slide into extinction will turn the Anthropocene into the Eremocene, the Age of Loneliness.

In spite of all their efforts, scientists have scarcely begun to understand the wildlands, as I have emphasized in these essays. We don't even know the size of the biodiversity that still survives. We think—we guess—that perhaps fifty thousand or even a hundred thousand more species of plants remain to be discovered, along with at least two million more of beetles, moths, wasps, and other insects. Of bacteria and viruses, we can only speculate how many kinds exist, and then to an order of magnitude. We wonder what that will be. A million? Ten million? A hundred million?

The biosphere of which all are a part is hierarchical. Select a human being for brief contemplation. Let it be you, standing, say, below the Murombodzi Falls on Mount Gorongosa—an appropriate location, since the birthplace of *Homo sapiens* included this region of Africa. Our species has been the most studied of all of Earth's species, yet your human organism is still very little known at the other levels of the hierarchy of which it is a part, including the ecosystems on which the organism depends, the species composing the ecosystems, the populations of each species, the organs of the organismic body, the tissues, and then the molecules composing your body's trillion or so cells. And the array does not stop there. Spread through your body are as many as a hundred trillion bacterial cells—about ten to each human cell. This immense microbiome lives in symbiosis with you as the host, its DNA flourishing alongside human DNA. Hundreds of the human-friendly bacteria have been DNA-sequenced and distinguished by microbiologists; it is likely that hundreds or even thousands more remain to be discovered.

Throughout the history of life, from its origin 3.5 billion years ago to this moment of you in Gorongosa, each layer in the biological hierarchy has changed in concert with all the others. A human exists in one spot within one ecosystem out of many, whether a mountain rain forest, a floodplain woodland, the edge of a lake, or a tributary stream. At Gorongosa these habitats harbor a variety of plants, animals, and microorganisms, and not just random species, either, but only those that can

Lake Urema is the heart of the Gorongosa floodplains—in the rainy season it floods an area of more than two hundred square kilometers (seventy-seven square miles), rejuvenating parched grasslands and allowing herds of herbivores to thrive.

live there in concert with the other species in this particular place. Of the dynamism of the Gorongosa ecosystems, of its food webs, energy flow, and nutrient cycles, we still know very little. We speak of them in very general terms and scratch out impressive-looking but mostly inadequate mathematical equations. It is mostly an illusion. The data are simply not yet there to do any better.

I mean for this confession of ignorance to be a sincere gesture of personal humility, and further an illustration of the dangerous relation of our species to nature. In one important respect, the living world of nature is vastly superior to humanity. Please don't jump to a misinterpretation, but listen to the argument further. The complexity of nature's layered

●●●

Sunset over the grasslands
of Gorongosa.

structure and the billion-year history of its construction may never be unraveled or copied. A warehouse of parallel zetabyte computers cannot simulate it with any degree of confidence. The starting conditions might be guessed, even duplicated in the laboratory. But the magnitude of the events in all the lives of the species, reverberating back and forth across the levels of biological organization from macromolecule to ecosystem, guided by natural selection through an all-but-infinite maze of possibilities, is beyond the comprehension that our still mostly Paleolithic brain, however sophisticated the culture, is yielding.

In striving to solve such daunting problems, an overreach of technology may be more of a detriment than a help. Some writers believe that relatively simple shortcuts are possible, that equal or better conservation might even be achieved with smaller amounts of fully protected land. One such nostrum, which might be called "Return to the Garden of Eden," is especially beguiling. It posits that nature has long since lost her virginity. Humans have trod upon every square foot of land surface able to support plants and animals. The evidence of our activity is everywhere, the argument continues, even if only in traces of manufactured chemicals. There are no more true wildernesses. To maintain biodiversity as humanity first encountered it is a delusion. It would be best, the Garden of Eden argument concludes, to find a way for humans and the surviving remainder of wild plants and animals to live together harmoniously. We should treat the living world as a magnificent garden. Wild plants and animals can live among people rather than the opposite, people among wild plants and animals. There the species would find protection, not in large sanctuaries reserved exclusively for them but among an enlightened people who love and look after them.

Perhaps I've drawn this image too starkly. But even if a mild and cautious version of such a philosophy were adopted, it would result not just in the end of wild ecosystems but also the extinction of a large part of Earth's remaining wild species.

If I read our own species right, humanity is too ignorant and selfish, and thereby inevitably lethal to most other species, for anything but large sanctuaries to work. The ecological footprint of a single human—defined as the amount of land each person requires for the necessities of food, water, habitation, transport, social activity, governance, and other purposes—is in excess of two hectares (about five acres). If the person has a habitation elsewhere, even in a faraway village or city, the footprint may be divided finely and spread widely, but it stays the same. In fact, it increases in the modern economy (think of distant transport and communication). If just one person's footprint is distributed to any degree into the world's biodiversity reserves, it will have a measurable effect on the fauna and flora.

A very few societies have evolved to live inside biological reserves with a small footprint, carefully designed to be sustainable. One is the Guarani people living in the northern Argentine state of Misiones. No species of plants and animals are known to have disappeared during their millennium-long residence. Although this achievement is admirable, modern peoples cannot even begin to live in the original Guarani way. It would be painful to endure for even a week. The Guarani way requires an encyclopedic knowledge of the resident plants and larger animals, and all the useful products that can be sustainably drawn from them. It demands care and even management of each species in turn to avoid its extinction. Another necessity is population control, whether by contraception or emigration. This the Guarani have achieved. And then of course, to stay pure Guarani, the economy must use a minimum of tools and medicinal products manufactured elsewhere, no matter how superior to native products.

Can biodiversity reserves be maintained if people are given free access to them? Can everyone explore and live within the reserves as they choose, however briefly? The answer is no. The result of such latitude of human impact on the fauna and flora, even if contained within strict rules of behavior, would be fatal to a large fraction of the species.

The closest sustainable approach to the use of natural area as precious gardens is by restricted visits to certain tracts and at certain times. Such is the practice of wildlife safaris of the better-managed African parks. At Gorongosa the tours away from Chitengo Camp are made over a small network of dirt roads, during the day and in the company of a guide or ranger. There is no cutting away of vegetation from the sides of the roads. A comparable control is used on Lignumvitae Key in the central Florida Keys, which harbors the largest West Indian lowland hardwood forest in North America—and, for that matter, almost all of the West Indies as well. The little island, only 112 hectares (276 acres) in area, and which I helped conserve, is strictly protected as the Lignumvitae Key Botanical State Park. A tourist trail runs through part of it, but the remainder is inviolate, with access granted only to the researchers and inspectors for the purpose of monitoring the condition of its environment and its biodiversity.

Of course, for all ecosystems everywhere there is the reality of climate change. Even if existing reserves are protected at the needed level on a local basis, ongoing climate change will increasingly threaten the species within them. A disturbing example is already apparent in the rich faunas and floras of the upper mountain forests of the Central and South American cordilleras. Their required environment is being changed by drying and by seasonal rainfall change. Being near the tops of the mountains already, they have nowhere else to spread. The same is true of the unique flora of South Africa: forced southward, its natural refuge stops abruptly at the Cape of Good Hope and Cape Agulhas. A similar and even more immediate peril faces the unique plants and animals of the high Arctic.

•••

Inselbergs in the
western part of
Gorongosa National
Park.

Most national parks and protected areas are today under another kind of threat: they are both too small and too geographically isolated by human occupation of the regions between them. Plant and animal species found in such reserves on the land can survive if their populations are able to expand into adjacent areas where temperature and rainfall regimes are changing in a direction more favorable to them. But animals and plants will not be able to make this necessary shift if urban centers, farmland, and cattle ranges completely replace habitats in between that might serve as stepping-stones to larger protected areas.

The solution to this dilemma is to connect the reserves with one another to create wildland corridors. Biologists and conservationists have begun to map exactly where such corridors are needed and might be possible. The most celebrated and immediately feasible would run north–south from the Yukon to Yellowstone National Park. The "Y2Y" corridor could be connected at the south to a second corridor of natural environments snaking down the Rocky Mountains. A continuation of the chain might then be made in the Sky Island mountains of southern Arizona, and thence across land of varying elevation to the Sierra Madre Occidental and Sierra Madre del Sur of Mexico. From there it would be possible to reach a corridor in the cordilleras of Central America.

These wildlife corridors, which could also serve as natural parks and wildlife reserves, might be constricted in places to as little as five kilometers (3.1 miles) across. The state of Florida is currently piecing together such a corridor that runs continuously from the Everglades National Park, at the southern end of the peninsula, to the Okefenokee National Wildlife Refuge, in Georgia, just across the Florida border.

It is both good conservation science and, I believe, practicable land management policy to envision corridors that "box" most of North America in segments running both north–south and east–west. From the top of a Yukon-to Mexico corridor, the west-to-east corridor in the north would be a segment within the coniferous forests and meadowlands of central and northern Canada. Another side of the box would run down the eastern United States from the Adirondack Forest Preserve through the Appalachian Mountains. The last segment of the box, now being planned by conservationists in Florida and Alabama (of whom I am proud to be one), would be the Gulf coastal corridor, ranging from the St. Marks National Wildlife Refuge and Apalachicola National Forest to Louisiana.

It may at first seem radical, but only this level of care applied to large sectors of the land and sea throughout the world will save global biodiversity. In order to ensure its permanent acceptance, it must also be designed to improve the quality of human life everywhere. How to serve both humanity and the rest of life is the great challenge of the modern era.

That is the reality of the alternative world we are trying to save in Gorongosa and other national parks and reserves around the world. There final sanctuaries are our transcendent heritage, and we will be wise to hold on to them. We can enjoy surviving fragments of Nature in various ways and measures. Let us first of all take constant pleasure from the surprise, mystery, awe, wholeness, relief, and redemption they offer. Deeper still, let us also hold on to a sense of the eternal latent in the wildlands. They provide hope for the immortality of life as a whole, freed of human cares and intervention and allowed to evolve as it did before we arose in Africa. While our species continues to manufacture its radically different and untested all-human world, the rest of life should be allowed to endure, for our own safety. While preserving our own deep history, it will, if we choose to let it, continue on its own trajectory through evolutionary time. By thus maintaining two parallel worlds on the planet, humanity will ensure the survival and continued advance of the rest of life, and of ourselves.

ACKNOWLEDGMENTS

I am grateful to many generous friends and colleagues who aided me in my visits to Gorongosa National Park, including especially Gregory Carr, Marc Stalmans, Kathleen Horton, Robert Poole, Piotr Naskrecki, and Tonga Torcida.

NOTES

2. Once There Were Giants

1. "Once There Were Giants," in *Sena-Gorongozi*, stories and poems assembled by Virlana Tkacz, Domingos Muala, and Wanda Phipps (New York: Carr Foundation, 2010).

3. War and Redemption

1. Philip Gourevitch, "The monkey and the fish," *New Yorker*, December 21, 2009, pp. 98–106, 108–11.

INDEX